MAKING SCIENCE EDUCATION RELEVANT

Kogan Page Books for Teachers series
Series Editor: Tom Marjoram

Fund-raising for Schools Gerry Gorman
Making Science Education Relevant Douglas P Newton
The Place of Physical Education in Schools Len Almond
Starting to Teach Anthony D Smith
Teaching Able Children Tom Marjoram
Towards the National Curriculum W S Fowler

MAKING SCIENCE EDUCATION RELEVANT

Douglas P Newton

Books for Teachers

Series Editor: Tom Marjoram

First published in 1988 by
Kogan Page Ltd
120 Pentonville Rd, London N1 9JN

British Library Cataloguing in Publication Data
Newton, Douglas P.
Making science education relevant.
1. Great Britain. Schools. Curriculum subjects: Science. Teaching
I. Title
507'.1

ISBN 1 85091 661 6

Printed and bound in Great Britain by
Billing and Sons Ltd, Worcester

CONTENTS

PREFACE

That science education should be both relevant and seen to be relevant is now a part of the dogma of science teaching. *Science 5-16: A Statement of Policy, Better Schools* (both Department of Education and Science, 1985) and *Better Science* (Association for Science Education, 1987) each reiterated it. The GCSE science syllabuses demand that teaching be relevant: this also figures largely in the National Curriculum recommendations for science education. The problem is that few people seem clear about what this means. What *is* relevant science education? How do we make our science teaching relevant? What it comes down to is: what are we trying to do when we teach science? The answer must be that we are trying to prepare children for their lives. A relevant science education, therefore, is something which contributes to that preparation. Explicitly relevant science education is readily perceived to do so by its recipients. I believe that the concept of relevant science teaching is not a simple one; it is not simply the inclusion of a few so-called 'applications' of science or discussion about social, economic and political issues. Science is in the curriculum because it *is* relevant and, it should always be added, relevant to *people*. Relevance is the very reason for its existence, and it should be the very backbone of science teaching.

For too long, we have tended to leave that relevance unstated: implicit statements are often no statements at all. The history of science teaching has been largely an account of conflict between the products and the processes of science. Are the theories, laws and generalizations of science to be paramount? Or should science education concern itself with how they are obtained? In practice, it matters little which wins or what the nature of the truce is if the result is seen as impressive only in its aridity. Science is not merely the sum of its products and its processes: it is products, processes and people.

This book is intended as an aid for those who teach science in schools to make their teaching relevant to those they teach. It provides a framework for use in designing and evaluating science courses. The framework is interpreted in terms of, and illustrated by, reference to science education at various levels of teaching.

I have followed the traditional practice of assigning gender by 'he' when referring to a teacher or child. This is not intended to imply that science is the preserve of men, but to avoid monstrosities like s/he, or worse, it, which have both found their way into some recent books on education. My concern that science should be made relevant to everyone is evident in the following pages. My hope is that this book will encourage more teachers to cross the theory–practice divide.

CHAPTER 1

VITALIZING SCIENCE EDUCATION

Introduction: products, processes and people

Science is more than its products and its processes. It is a human activity which affects us all. The products of science – concepts, laws, principles, generalizations, theories and models – will not let themselves be ignored. They affect our relationship to the world and our place in it. Whether or not someone knows of Newton's Law of Universal Gravitation or Darwin's Theory of Evolution matters little; he will absorb a world view which has been moulded by such theories. At the same time, the skills and processes associated with science – observing, classifying, measuring, hypothesizing and experimenting, for example – will be perceived as successful and powerful instruments of investigation. As a consequence, what is often described as the 'scientific method' is used in areas far removed from science. Even history, the Queen of Humanities, has been tempted by it (Aydelotte, 1971).

In school, technology is often linked with science. This may be because they are associated with practical utility and applications. Indeed, for the primary school, the National Curriculum Working Group describes science and technology as being 'very closely interrelated', although it does recognize that 'technology finds its expression in activities across the curriculum and not just in science-based activities' (DES, 1987).

The relationship between science and technology is not simple. Although they may not share intent, they are seen as sharing products and processes. For example, physicists and engineers might both try to understand the behaviour of the nucleus. Both would consider scientific generalizations about it (products of science) as being worthy of attention, and they would probably use similar skills and processes to do so. But one might hope to develop a theory about nuclear structure while the other might want to design a better nuclear reactor. In practice, even differences in intent are unlikely to be clear-cut. Is it unknown for a physicist to make something useful or for an engineer to theorize? Consequently, aspects of technology are usually seen as having a legitimate place in a school science lesson.

The impact of technology on our lives is both substantial and complex. We live the kind of life we do by courtesy of technology. Large sections of the population are so dependent on it that they would perish if denied its support (Dessel, 1973). It seems likely that science and technology will influence our lives more and more as they now give us our main means of controlling and changing our environment (Beveridge, 1980).

Science education has concerned itself almost exclusively with the products and processes of science and with the balance between them. This is not enough. Science is made by *people* and this has tended to be ignored, perhaps even suppressed. Referring to physics teaching, Woolnough expressed the opinion:

> We need to present a more human view of physics, to recognize that in seeking to understand and to solve problems in the physical world we are dependent on creativity, tacit knowledge and value judgements as well as logic, analysis and fundamental principles. (Woolnough, 1986)

This criticism applies equally to chemistry, biology and other sciences.

At the same time, science, even the purest, is made *for* people. For the scientist, it offers a satisfaction of personal needs. To society (and also to the scientist) it offers a comprehensible, ordered view of the world that offers some hope of controlling and manipulating it. The effective control and manipulation of nature has the potential to affect dramatically all life on this planet, for good or ill. The educational implication has been highlighted by Sir George Porter: 'Science makes change possible – everyone needs to be in a fit state to judge whether we want that change' (Porter, 1986).

Science education should not ignore the relationships between science, technology and mankind. It is in these relationships that the relevance of science lies. Unless the nature of that relevance is made evident, a totally inadequate view of science and its attitudes will prevail in a world where science is pre-eminent. In such a world, science education must reflect a trinity of concerns: *products, processes* and *people*. When it does so, it becomes a humanized science education and it stands a chance of being relevant.

Humanized science teaching

The feeling that a science education should do more than feed children a cold diet of the products and processes of science is neither recent nor completely novel. Champions of science education have always emphasized a wider relevance, beyond that of mere utility. Of course, their motives over the decades have been different, since they spring largely from the perceived needs of science education and society at a given time.

At the beginning of this century, there was concern that national interests were suffering through an ignorance of science. A *Neglect of Science* committee was formed to press for greater provision for science and science teaching. Classical scholars responded with a vigorous defence of their educational interests. They argued that while science might make someone more rational, it could not make him more human. In the face of this, the defence of science education solely on the grounds of its practical utility and as a hone for sharpening the faculty of reasoning was inadequate. If science was to be afforded a firmer footing in the curriculum, science education must be seen as able to humanize – in short, to make better people.

The Thomson Report of 1918 said that science would be a humanizing influence if its 'human interest' was developed side by side with its 'material and mechanical aspects'. R A Gregory, the influential advocate of the 'Science for All' movement, held the view that 'school instruction in science is not...intended to prepare for vocations, but to equip pupils for life as it is and as it soon may be' (Gregory, 1922). He said that science courses should be broadened and humanized by an historical method of teaching. Some science teachers were also reacting against the excesses of heurism and felt that science courses were too narrow and lacking in an overall and comprehensive view of science (Sherratt, 1982). It was felt that a concentration upon the 'experimental method' had led to a neglect of 'science as a body of inspiring principles and a truly humanizing influence. There should be more of the spirit and less of the valley of dry bones, if science is to be of living interest.' The means of redress was seen to be through teaching about the lives of scientists and the history of science (Gregory, 1917).

For some, humanized science teaching implied ancillary courses in the methods and philosophy of science. For others, it meant a less formal, even 'anecdotal', approach, such as that of E J Holmyard, in which the lives of the pioneers and the atmosphere of their discoveries were included as opportunities arose in science lessons. *General Science*, born of the 'Science for All' movement, required candidates to have some acquaintance with the historical background of science, but examination questions were few and limited to those of a biographical nature (Fowles, 1949).

While humanizing science teaching might be seen as a way of pointing out the cultural worth of science education, of establishing the respectability of science as a school subject, of broadening it, showing its relevance, importance and significance for people and of engendering strong interest, in practice, it meant including, in one way or another, some history of science. For John Brown, a London school inspector of the 1920s, the history, romance and biography of science were important components of science teaching if its cultural value was to be realized (Brown, 1925). HM Inspector, F W Westaway, in 1929, echoed this view: 'the history of scientific discovery opens up to the imagination great pictures of the work of great men, thereby placing science in the front rank of humanistic studies' (Westaway, 1929). His Majesty's Inspectors, in 1933, added that an aim of science teaching was to show science to be a 'scientific endeavour', which, according to W L Sumner, Lecturer in Education at Nottingham University, meant placing it 'in its temporal and historical context' (Sumner, 1936).

By the 1930s, other strands were being added to the aims of science teaching by politically minded scientists such as Lancelot Hogben. His book, *Science for the Citizen*, which he described as 'the first British handbook to Scientific Humanism', was intended to bring an understanding of science and the scientist to everyone so that an electorate could make well-informed decisions in a scientific society. Within this overall aim, there was the intention of showing science to be a human activity, full of human endeavour with intellectual satisfactions, as well as being a means of satisfying man's material needs. He also wanted to discredit the popular misconception that the scientist had no time to be socially responsible and to warn that science can have adverse social implications unless society is vigilant (Hogben, 1938 and 1942).

J D Bernal also emphasized the social aims. Science education, he wrote, must '...provide enough understanding of the place of science in society to enable the great majority that will not be actively engaged in scientific pursuits to collaborate intelligently with those that are and to be able to criticize or appreciate the effect of science on society'. For both Hogben and Bernal, this could be achieved by making the history of science 'a vital part of science teaching...' and by relating science more closely than before to 'the material and social aspects of ordinary life' (Bernal, 1946).

Most of these points were reiterated in the 1961 Policy Statement of the Association for Science Education (ASE): 'the view was expressed that "science should be recognized – and taught – as a major human activity which explores the realm of human experience..." and it should include "the effects of science and technology on human life" ' (Kerr, 1966). But, during the same decade, F R Jevons pointed out that science as it is taught is not the same as science as it is practised. The remedy, once again, was to use an historical approach, but tempered with restraint as, he says, the glamour of science is often over-estimated (Jevons, 1969). At about the same time, T S Kuhn's *The Structure of Scientific Revolutions* (1970) was published criticizing the image of science perpetuated by science teaching. Students and professionals, he claimed, came to feel like participants in a long-standing historical tradition which had never existed. Now, Cawthron and Rowell have gone as far as defining humanized science teaching as that which is intended to modify 'the traditional positivistic ethos of school science'. They would correct the image of science by including 'epistemological and socio-psychological components into our science curricula' (Cawthron and Rowell, 1978).

Science and General Education, published by the ASE in 1971, had nothing to say about this criticism. In part, it is almost a restatement of the Thomson Report of 1918 when it argues that science illustrates the perennial problems of the human situation just as well as other subjects which have been traditionally regarded as 'the humanities'. It added that science 'should be presented in ways which show how its applications influence the patterns of modern life and social organization'.

By 1979, the ASE's *Alternatives for Science Education* included amongst its aims an appreciation of man's cultural and historical predicament, an understanding of advanced technological societies and the cultural aspects of science. These aims were to be achieved by the 'inclusion of history, philosophy and social studies in science studies'. Subsequently, *Rethinking Science* (Passmore, 1984) described how social issues make science 'more interesting and enjoyable' and show it to be a human activity. Seven ways in which science might be shown to be relevant were outlined. These covered: the utility of science; science in the world of work of the scientist and technologist; an interdisciplinary approach; an historical approach; a sociological approach emphasizing a preparation for citizenship; environmental crises and controversial issues as starting points; and the philosophy of science.

More recently, the Department of Education and Science stated, in *Science 5–16* (DES, 1985a), that 'science education should be presented and assessed in a way that allows the pupils to see its direct relevance to their lives... One test for the inclusion of topics or approaches in a science course... should be their value to the pupils... in their adult and working lives in the world of the future'. In *Better Schools* (DES, 1985b), it is said that 'The principle of relevance means that teachers should be skilful in drawing on pupils' experience and helping them to apply what is learnt to life outside school, so preparing them more effectively for working life.' Here, it seems that a utilitarian view of relevance is being taken.

The GCSE science syllabuses must be in accord with National Criteria for each subject which, in turn, must agree with General Criteria. The General Criteria 'require that all syllabuses should be designed to help candidates to understand the relationship of the subject to other areas of study and its relevance to candidates' own lives' (Secondary Examinations Council, 1986). This has often been interpreted as developing an informed interest in scientific matters and becoming confident citizens in a technological world while considering the social, economic, technological, environmental, ethical and cultural influences and limitations of science. However, at the insistence of the DES, the majority of the 15 per cent of the marks allocated to these areas is to go to technological applications (Gibson, 1986). In effect, the pupils are to be shown how to drive the science-technology machine but they are not to look where it is going.

The National Curriculum Working Group for Science was obviously influenced by this background (DES, 1987). The members emphasized the need for relevance to each pupil's everyday experience, to personal and public health, and to today's world. There should be, they say, an awareness of practical applications and an appreciation of economic, social, personal and ethical implications of science. They add that children 'should be encouraged to explore some of the moral dilemmas that scientific discoveries can cause', should learn to 'appreciate their responsibility as members of society' and have responsible attitudes to living organisms and the physical environment. Their aim is for children to 'gain balanced insights into the importance of science and technology in the economy and to the quality of life of all citizens'.

Although it has been felt that a science education might do more than offer a knowledge of products and some practice in skills and processes, what was expected of it in this respect has not always been the same. In the early decades, school science had to be seen as being able to educate. This meant it had to help to make human beings out of the somewhat baser clay in the classroom. This was to be achieved by teaching *about* science and, in particular, about the history of science and the lives of scientists. In practice, however, these elements did not force themselves strongly into the prescribed curriculum. Examination boards were mainly concerned to assess a knowledge of the products of science. Later, the emphasis was less on improving the individual as a person and more on making him better equipped for his role as citizen. However, the Second World War and the manpower demands of succeeding decades promoted specialization which tended to exclude such luxuries (Waring, 1979).

Recent decades have seen a concern with the image of science. The mismatch between the image that people have and perceived reality seems too large to accept and humanized science teaching is expected to reduce it.

At times, teachers like Holmyard have valued humanized science for the interest it engendered and, presumably, for any resulting enhancement of learning. As a teaching method, it continues to have supporters. For example, Ebison, in a report of a meeting held at the Royal Institution in 1984 'spoke of the role that history might play in improving physics teaching. For one thing it could help humanize physics for students'. Others were 'particularly interested in exploring the social origins of certain scientific developments' and of using historical material as 'an effective way of teaching science' (Meadows, 1984).

Aims have a way of accreting, (Uzzell, 1978) and it would be truer to say that those of humanized science teaching have widened rather than changed, although at different times some have received more attention than others. For much of the time, it was largely through the history of science and biographies of scientists that these aims were to be achieved. In recent years, other ways of achieving them have been explored. *STS* [Science and Technology in Society] *for Schoolchildren* begins with an historical study, then looks at how the topics affect us today.

Pupils are encouraged to contribute their own ideas about society's needs and values (Solomon, 1981). In a similar vein, but for use in sixth form General Studies courses, are the booklets of the Science in Society (Association for Science Education, 1981) and the Science in a Social Context (SISCON) projects (Solomon, 1983). A recent development is the Science and Technology in Society (SATIS) project which aims to relate science to its social and technological context (Holman, 1986). It uses, for example, controversial issues, group discussions, role-play, problem-solving and reading activities in an attempt to show the relevance of science. In doing so, of course, it humanizes science education (Newton, 1986a).

Wider aims

Although official statements have mentioned broader and humanizing aims for almost a century, they have, in general, been afforded only lip-service. One reason may be the abstract and diffuse way in which such aims are often expressed. What these aims were expected to achieve can be divided into four main groups: moral, contextual, philosophical and epistemological, and psychological.

One aim of some who would humanize science teaching is directed towards *moral education*. Science teachers should not be surprised by this. Aspects of morality are manifest in almost any way of teaching science. We have a concern for our discipline's standards, we try to inculcate a respect for truth, accuracy, impartiality and objectivity as we see them, and we hope to encourage a willingness to listen to others and to be subject to evidence and reason. However, when the aim is a humanizing one, its target is to make people rational *and* human. This extra step involves a process of decentring, a sensitivity to human nature and an awareness that other people have feelings and wills of their own (even scientists). These should lead to sympathetic responses in decisions and behaviour, so that rationality is tempered with humanity when considering the moral acceptability of choices.

Such aspects of moral education are often not seen as the legitimate province of school science teaching. Yet their neglect in the very discipline that is patently so powerful in the world is like a value judgement which invites the treatment of animate and inanimate without distinction.

Another intention is often to provide one or more *contextual settings* for a subject. A particularly powerful way of showing why science is the way it is is to trace aspects of its history. Within a broad historical perspective are many possible settings, ranging from a simple ordering of events in a chronology to the location of those events in the history of the growth and change of ideas. The American Project Physics, for example, does both of these amongst other things, and gives what might otherwise be isolated areas of knowledge an overall coherence. It embeds the subject in a broader world picture.

Settings are seldom mutually exclusive. While one might be emphasized, others can be drawn in. The history of ideas, for example, can touch on aspects of psychology in considering the unique contributions of individuals. It may also bring aspects of sociology into play when, for example, discussing why a particular discovery was made when it was. The historical perspective itself might be replaced by one which sets science against the recent and immediate background, such as the impact of technology on society. Science and technology affect culture (in the sociological sense) and give rise to palpable change in the way we live. Their impact on society has been used to provide settings in such projects as *Science and Technology in Society*, *Physics Plus* and the *Science Support Series*. The Secondary Science Curriculum Review, in *Better Science*, described relevant science education in terms of such contexts (Stewart, 1987). On the other hand, a simple historical setting or context has been used for younger children with biographies designed to show scientists as real people rather than extraordinary paragons (Newton and Newton, 1986).

Underlying the provision of contextual settings there often seems to be a belief in the value of coherence and unity of knowledge for rational action. Action in one sphere of human activity often has ramifications elsewhere. The coherence and unity provided by contextual settings can make this apparent.

Humanized science teaching has also been advocated to correct what are seen as inappropriate images of science and what the scientist is trying to do. For example, science has been taught as an accumulating wisdom resulting from a continuous mining of nature. Each generation of scientists adds its contribution to that of its predecessors. Supporters of Kuhn's theory of paradigm science and revolution argue that this image is inadequate and it has been proposed that historical case studies might be used to modify it. Humanized science teaching would then be a means of acquiring a view of science and scientific method which is in accord with the accepted *philosophical* and *epistemological* theories.

While humanized science teaching might indeed be used to achieve such aims as these, it has also been advocated for its *psychological* advantages as a teaching method. Contextual settings, for example, allow knowledge to be woven firmly into the broader backcloth with, presumably, improved mental schemata, enhanced retention and increased potential for the articulation and application of that knowledge. Biographies, on the other hand, may provide opportunities for emulation, role play and fantasies of achievement (especially with younger children), with the possibility of enhanced motivation, rehearsal and integration of knowledge. Holmyard's and McKenzie's Historical Method of teaching chemistry and physics, respectively, was to develop topics in accordance with their history, taking advantage of the human interest value of the participants in those historical settings. Their aim was to generate interest and positive motivation and, as a consequence, to enhance understanding and learning.

Research suggests that positive attitudes towards science, especially amongst girls, are more likely to be encouraged by a consideration of the social implications of the applications of science. The humanized treatment of science which effects this is then a vehicle similar in intent to the Historical Method.

Now, although these four groups of aims are often expected to be achieved by humanized science teaching, they differ in their intention to *humanize*. Some aspects of the moral aims, such as a willingness to temper rationality with humanity, are clearly humanizing in intent. Similarly, taking a broader view provided by some contextual settings may also mean considering moral dilemmas and making value judgements which would affect people. The philosophical aims, on the other hand, can be more concerned with inculcating an acceptable image of the nature of science and its methods. Similarly less concerned to humanize are the psychological aims, which are focused on enhancing interest and learning in connection with products and processes in science. Any wider aims achieved would be a bonus, although probably not an unwelcome one.

In practice, humanizing science for philosophical or pyschological ends can provide an opportunity for some achievement of humanizing aims. Equally, where humanizing aims have priority, the material used to achieve them may, by its nature, also modify images, enhance interest, motivation and learning (Newton, 1986b). Of the four groups, it is probably the moral aims which are the most neglected, but even the contributions in the other areas by the various Science, Technology and Society enthusiasts have yet to spread in any fundamental way into the core of school science teaching. At the secondary level, there has been increased comment on these wider aims (Fig. 1) but it has had little impact on textbooks for the O- and A-levels of the GCE and those for the GCSE seem to have the same flavour (Newton, 1986c). The educational aims of successful textbook writers tend to be focused on products, to some extent processes, and mainly on passing examinations (Newton, 1983). This does not mean that these textbooks are bad. Presumably, they give teachers largely what they are used to and are comfortable and familiar with, and they help their readers through examinations. But, in not giving some attention to *people*, there remains a lack of exemplar material of a kind which is integrated with the products and processes. For the teacher, it remains unclear how science teaching can be made relevant without prejudice to the other aims of science teaching which must remain at the centre of attention. This helps to perpetuate the *status quo*.

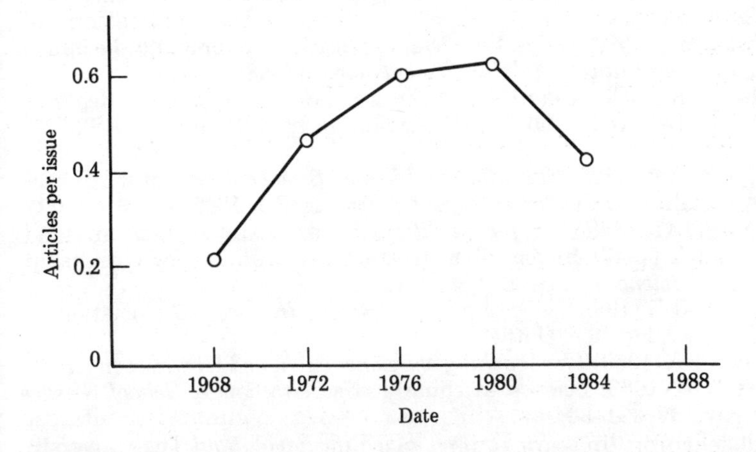

Figure 1.1 *Mean number of articles relating to humanized science teaching per issue of the 'Journal of Biological Education', 'Education in Chemistry' and 'Physics Education' combined as one. The mean was found for each four year interval beginning with 1967 and ending with 1986. Issues devoted solely to careers were excluded.*

References

Association for Science Education (1979) *Alternatives for Science Education* ASE, Hatfield.
Association for Science Education (1981) *Science in Society Readers: Books A to L* Heinemann, London.
Aydelotte, W O (1971) *Quantification in History* Addison Wesley, Reading, Ma.
Bernal, J D (1946) Science teaching in general education. *School Science Review* **27**, 150–158.
Beveridge, W B (1980) *Seeds of Discovery* Heinemann, London.
Brown, J (1925) *Teaching Science in Schools* University of London Press.
Cawthron, E R; Rowell, J A (1978) Epistemology and science education. *Studies in Science Education* **5**, 31–59.
Department of Education and Science (1985a) *Science 5–16: A Statement of Policy* HMSO, London.
Department of Education and Science (1985b) *Better Schools* HMSO, London.
Department of Education and Science (1987) *The National Curriculum Interim Report: Science* HMSO, London.
Dessel, N F; Nehrich, R B; Voran, G I (1973) *Science and Human Destiny* McGraw-Hill, New York.
Fowles, G (1949) The place of the history of science in education. *School Science Review* **31**, 2–6.
Gibson, M (1986) Physics: the social implications. Examining the issues. *The Times Educational Supplement*, **3663**, 47–48.
Gregory, R A (Chairman) (1917) *Science in Secondary Schools* Report of the British Association for the Advancement of Science (BAAS), 123–207.
Gregory, R A (1922) *Education and School Science* Report of the British Association for the Advancement of Science (BAAS), 204–208.
Hogben, L (1938) *Science for the Citizen* Allen and Unwin, London.
Hogben, L (1942) Biological instruction and training for citizenship. *School Science Review* **23**, 263–281.
Holman, J S (1986) *Science and Technology in Society* Association for Science Education, Hatfield.
Jevons, F R (1969) *The Teaching of Science* Allen and Unwin, London.
Kerr, J F (1966) Science teaching and social change. *School Science Review* **47**, 301–304.
Kuhn, T S (1970) *The Structure of Scientific Revolutions* The University of Chicago Press, Chicago.
Meadows, A J (Chairman) (1984), Report of a meeting held at the Royal Institution on 1 February 1984 to discuss the possibility of setting up a History of Physics group within the Institute of Physics.
Newton, D P (1983) The sixth form physics textbook. *Physics Education* **18**, part 1, 192–198, part 2, 240–246.
Newton, D P (1986a) Humanised science teaching – what is it? *School Science Review* **67**, 457–461.
Newton, D P (1986b) Products, processes and people. *School Science Review* **68**, 350–355.

Newton, D P (1986c) Humanised science teaching and school science textbooks. *Educational Studies* **12**, 3–15.
Newton, D P; Newton, L D (1986) Humanising primary science. *Education, 3–13,* **14**, 47–52.
Passmore, E L (1984) *Rethinking Science* Association for Science Education, Hatfield.
Physics Plus (1985 etc) A curriculum project sponsored by the Standing Conference on Schools' Science and Technology, Cambridge, Hobson's Ltd.
Porter, Sir G (1986) What is science for? *New Scientist* **1535**, 32–34.
Science Support Series (1983 etc) A collection of topic booklets for sixth-form physics teaching, Hobson's Ltd, Cambridge.
Secondary Examinations Council in collaboration with the Open University (1986) *Science GCSE, a guide for teachers* Open University, Milton Keynes.
Sherratt, W J (1982) History of science in the science curriculum: an historical perspective. *School Science Review* **64**, 225–236.
Solomon, J (1981) STS for schoolchildren. *New Scientist* **89**, 77–79.
Solomon, J (1983) *Science in a Social Context* Association for Science Education, Hatfield.
Sumner, W L (1936) *The teaching of science* 4th edn, Basil Blackwell, Oxford.
Thomson, J J (Chairman) (1918) *Natural Science in Education* Report of the Committee on the position of natural science in the educational system of Great Britain, HMSO, London.
Uzzell, P (1978) The changing aims of science teaching. *School Science Review* **60**, 7–20.
Waring, M (1979) *Social Pressures and Curriculum Innovation* Methuen, London.
Westaway, F W (1929) *Science Teaching*, Blackie, London.
Woolnough, B E (1986) Physics as a human activity. *Physics Education* **21**, 1–2.

CHAPTER 2

THE RELEVANCE OF SCIENCE

Potential to fulfil needs

The difficulty with the moral, contextual, philosophical and epistemological groups of aims described in Chapter 1 is that they seem to comprise disparate responses to the needs of people and science education as they are perceived to be at a given time. A need is detected and the feeling that something should be done about it is expressed as an aim. As life changes, needs change. Old aims fade (but seldom die) and new ones are born in response to new needs. At any one time, the bag of aims bulges with its motley contents. However appropriate at that time, the shape of the bag is no help as needs change. The only response it can make is to let old aims sink to the bottom and open its neck to new. What is needed is a framework for the aims so that it makes their focus of concern apparent. In doing so, it should help us to identify areas of potential weakness and to consider where appropriate emphases should lie. What humanized science teaching is expected to achieve may change with time, but the framework could provide a coherence which is more enduring. The lumps in the bag would no longer be perceived as lumps, but as legitimate emphases within a broader structure.

One way of arriving at such a structure is to ask why science is important. Science has been described as 'the most distinctive enterprise of Western civilisation in the twentieth century' (Broad and Wade, 1983), and like many, perhaps most, human activities, science has the potential to fulfil needs. There are three main ways in which science might do this:

- as a human activity
- as a model for investigation and problem solving
- as a world view or view of reality.

Each has moral and philosophical aspects and each has consequences, some of which are good, some bad and some of no account. The relevance of science to the lives of each and every one of us can be traced back to this potential. It is in the attempt to realize this potential, to fulfil such needs, that we find the human dimensions of science. Ultimately, the importance of science lies in what it has to offer us. What is on offer is now described in more detail.

Science as a human activity

For the individual, science offers the fulfilment of important, personal needs. Boyle wrote that it was 'not without delight and satisfaction' that he confirmed his thoughts 'touching on the Spring and weight Air'. It must have been singularly without these feelings that Dr Oldham opened *Nature* on December 1st, 1870, to read:

'*The Times of India* states that the coal-beds discovered in Bellary are excellent in quality and abundant in quantity. Great anxiety is felt for Dr Oldham, who said he would eat all the coal found in the Madras Presidency, for the doctor is a man of honour.'

Science is like that. Sometimes there is delight and satisfaction, sometimes frustration, disappointment and even mortification. Whether the need is of approbation, beauty, competence, opportunity for creativity, comprehension, satisfaction of curiosity, unity of understanding, or whatever, its fulfilment – or lack of it – leads to some emotional response. The activity owes its birth and continued existence to the satisfaction of needs – human needs (Newton, 1986a). In common with many other human activities, science can require moral and ethical behaviour. For example, there is some expectation that there will be a regard for ideals of honesty and self-suppression. A flagrant disregard of such ideals can, on occasions, have serious consequences. While we might be amused by the claim that: 'The reason that the Govt. has refused to aid the expedition to observe the approaching eclipse is that it is perfectly assured that men of science and culture are nothing but a set of lying imposters, and would swindle the public out of thousands of pounds...' (*Nature*, October 20, 1870), instances of deception are not unknown. These range from an unconscious manipulation of data, reflecting a tendency to see what the scientist wants to see, to outright fraud. Self-deception, as with Lovell's canals on Mars and Blondlot's N-rays may be endemic in science. On the other hand, scientists, being people, feel needs to different degrees. For some, the need for approbation, fame and its rewards is strong enough to override ethical and moral obligations. Cyril Burt's fabrication of data to support his theory of the inheritance of intelligence is well known. It seems that Robert Millikan tended to report only favourable results, and the feasibility of some of John Dalton's experiments is open to question. Broad and Wade describe some 34 such instances taken from classical to recent times (Broad and Wade, 1983).

In science, there is a need to persuade others to accept evidence, theories and arguments, so the ideals of honesty, integrity, self-suppression and open-mindedness shape the way in which science is reported. Science is purged of its human origins (its Achilles' heel) before publication. Textbooks perpetuate and propagate the consequent dehumanized image of science so that popularly, scientists become calculating automatons, unfeeling and unaffected by matters emotional and aesthetic. Add to this the perceived difficulty and abstract nature of science and a dose of suspicion arising from ignorance, and we should not be surprised to find images of the scientist which range from explorer, sage and altruist to nutcase and arch-villain (Hales, 1982). But science is not an activity for Philistines *per se*. Many scientists have been poets, novelists, musicians and artists (Goran, 1974). Neither are the opinions of scientists as infallible as even the humble pocket calculator. 'I have not the smallest molecule of faith in aerial navigation other than ballooning' wrote Lord Kelvin (Beveridge, 1980). Nor do they all maintain an indifference to the uses made of the products of their labour: Faraday would have nothing to do with poison gas. They do not confine their thoughts to their laboratories. In 1968, 2000 scientists in the USA signed a document calling the birth control encyclical of Pope Paul VI repugnant on the grounds that it would cause the deaths of many people.

Scientists have their fears, hopes, and even personal conflicts, like everyone else. Witness the mutual dislike of Newton and Hooke: Newton would not accept the presidency of the Royal Society until Hooke was no longer associated with it. Davy's relationship with Faraday turned sour and he opposed Faraday's fellowship application to the Royal Society. The chemist, Frederick Soddy, was bitter about his lack of recognition for his work with Rutherford on radioactive disintegration theory. All is not harmony between scientists.

Scientists are not so stable that they are immune to overwork and, amongst other things, lack of recognition. Newton, Marie Curie, Maxwell and Faraday suffered from some degree of mental illness for periods of their lives. Others have taken their own lives when under stress. Paul Kammerer, a biologist, killed himself partly because his experiments on inheritance were reported to be fraudulent (Goran, 1974).

In view of these observations, it needs to be said that scientists *are* people. For them, science fulfils needs. At the same time, whatever common traits might be found in scientists, the popular stereotype fails to do justice to their variety (Lendrem, 1985).

Science is not merely the activity of individuals. It is not practised by isolated scientists who work only to satisfy their personal curiosity. It is also a social activity. From a philosophical point of view, it cannot be said to be truly the product of any one person. According to Chalmers (1978), 'From the level of the results of individual experimenters and theoreticians right up to the level of the complex theoretical structures that constitute scientific knowledge, the activity and products of science transcend the actions, motives and beliefs of individual scientists... The skills of an experimenter are partly learned from textbooks but mainly by trial and error and interaction with more experienced colleagues... The same can be said of the purely theoretical [work]... [It involves] assumptions, mathematical techniques, approximations, etc.', which are largely harvested from the work of others. Modern physics theory, for example, thus cannot be identified clearly with the beliefs of any one physicist. In turn, no-one keeps his discoveries to himself – not necessarily through altruism, but in the hope of credit, recognition, approbation, acclaim, or simply promotion. The institutionalization of science imposed a sense of order and gave greater coherence to the sub-units of the 'scientific community'. As well as fulfilling a need for affiliation and professional status, the institutions provide windows for the display of the scientist's goods and a structure for their appraisal. It is often through these institutions that the scientific community affords laurels to its members, on a sufficiently parsimonious basis to signify exceptional merit.

Belonging to a group can also have very tangible benefits as far as a scientist's work is concerned. Nobel laureates tend to begat Nobel laureates, metaphorically speaking. For example, Liebig was at the head of a line of some thirty such laureates (Beveridge, 1980).

Thus science is a social activity and it cannot be fully understood out of its social context. Some philosophers, like Kuhn, afford this aspect a central place in the study of science. Others, like Chalmers, acknowledge its importance but give it a less central role.

Science as a model for investigation and problem solving

It is largely through technology that we adapt to the environment and adapt that environment to ourselves. It would be an over-simplification to ascribe this adaptation directly to the application of science (Hodson, 1985 and Taylor, 1973). Science is not always the mother of invention. Take, for example, the centripetal governor. Used to control the speed of millstones in the early nineteenth century and possibly based on a patent of 1787, its full, theoretical analysis was not available until 1868. On the other hand, neither is necessity always the mother of invention. Those who remember the invention of the laser may recall that it was described as a device looking for an application. It is more useful to think of pure and applied science as being at opposite ends of a continuum. Workers at different points on the continuum do not see their location as fixed and are not necessarily blind to the activities of others. But, whatever the differences in their ultimate goals, their approaches seem to share some common features. Mechanistic and reductionist in outlook, with attitudes of disinterest towards people, carefully controlling variables, supporting and testing ideas and theories with experience and experiment, they seem to be in possession of a powerful, effective and singularly successful way of ordering the natural world, both mentally and physically. That this has influenced attitudes and approaches outside science need hardly be said. From Evolution applied to societies and industries to time-and-motion work studies, we see the adoption of 'scientific' perspectives.

Science as a model for solving problems and for adapting to the environment, like science as a human, personal activity, gives rise to moral and philosophical issues. When we were obliged to let nature take its course, there was no choice and no moral issue. Now, science and technology have given us the power to deflect the inevitable and to change the course of nature. They offer us solutions and thereby generate the need for choice. For example, attention is currently focused on energy supply and problems of population and pollution. The choices made regarding these will very much affect the quality of life of future generations, for better or worse. Problems such as these are enormously complex and their ramifications extend into remote and unexpected, yet important, areas. There is a need to foresee the implications of action but, even if the consequences can be teased out and exposed, there still remains the bottom line: do we want to go there? Because satellite television is feasible, for example, is not to imply that it is desirable. There is a need to appreciate that there *is* a choice, that an action has consequences and that decisions should be rational and humane (Ullrich, 1983).

Surprisingly, although science has proved to be a successful tool for bringing about change, philosophers are by no means in accord about its nature or its methods. There seems to be a general agreement that a Baconian scientific method, based on induction, is unsound. A scientist does not collect lots of observations, stare at them for a while and then perceive the pattern which links them together. However, opinions regarding what does happen seem to range from the view that there is no such thing as scientific method, that science is anarchic and anything goes, to the view that a pattern can be seen in the orchestration of scientific processes. Consequently, the status of the products of science – its concepts, theories, generalizatons and laws, for example – are not absolutely clear. The choices which science presents should be seen in the light of this. Its processes are unlikely to be infallible and its products will not be immutable. Science should be presented in a way which reflects this but, at the same time, this should not be given more than its due weight. Whatever their limitations, science and technology may be the best tools available when it comes to understanding Nature and coping with the environment (Hodson, 1986).

Science as a world view or view of reality

Science codifies and organizes a body of knowledge into a world view. It offers us a mental map of Nature which locates mankind, collectively and individually, in relation to its parts and provides us with a sense of identity (Walsh, 1964). For example, when the Earth was considered to be at the centre of the universe, Man saw himself as the most important being in existence: everything was for him. The displacement of the Earth to the periphery of things diminished his stature. He became merely one amongst many and the many had as much right as he to a share of things.

The particular projection of the mental map – the principles used to draw it – owes much to people like Réné Descartes (1596–1650). Seeing Nature as a mechanical clock, he looked for an understanding of it in the function of its wheels, cogs, ratchets and pawls, a view which was encouraged by the sight of the intricate automatons which were becoming common at the time.

The popular amazement at these automatons is reflected in the legend that Descartes himself built a true-to-life android, Francine. The reality is that she was probably his illegitimate daughter.

Cartesian reductionism is now at the foundation of Western culture (Capra, 1982; Richardson and Boyle, 1979). The view is that 'there is nothing but physical and chemical laws and that, ultimately, all things can be explained in terms of physics and chemistry' (Beveridge, 1980). Descartes retained the Prime Mover in his view of Nature, and later scientists, like Linnaeus, saw their work as revealing that of this Prime Mover (Newton, 1986b). These were still anthropocentric world views. Man was the finest object of Creation and all Nature was subordinate to him. But, however fine that mechanism might be, its cogs and gears bear not a Creator's stamp but, according to Jacques Monod, merely the mark of blind chance (Monod, 1972). His work on genetic structure perhaps exemplifies the enormous power of a reductionist-mechanistic approach. Mankind's retreat from the centre of things began long before Monod expressed his views, but he may be the one who dealt the death blow. Such a world view has no room for appeals to a Higher Authority.

While acknowledging the success of the reductionist-mechanistic world view, others assert that the whole is not simply the sum of its parts. Only with an holistic view, they claim, can this organizational charisma be understood. Capra (1982) claims that a tendency to adopt such a view is now evident in the work of scientists and that the limitations of reductionism need to be more widely appreciated. Capra claims that a systems view of life is necessary for a fuller understanding of biology, that medicine is improved by integrating physical and psychological therapies and that quantum physics exemplifies an ecological approach to science. Monod, on the other hand, sees such a view as merely reflecting a perennial need to find intent and purpose in life when, in reality, there is none. Whatever the outcome of this debate, it is the reductionist-mechanistic world view which is popularly pre-eminent. But, more to the point, this is not just a paper pre-eminence. It tends to determine our global behaviour, that is, how we relate to, act in and respond to the world, both animate and inanimate. It has been suggested that the disease of modern society is alienation caught from a world view in which alienation is seen as the only means of achieving a valid relationship with reality (Roszak, 1968). Certainly, such a world view chooses to divorce itself from aesthetic, personal and moral considerations. Without a human and humane input, a depersonalized, dehumanized culture develops, lacking in human sympathy and indifferent to feelings.

Science education and the relevance of science

Science and technology do so much for us and to us that it is important that we are aware of their roles in our lives. Science education, however, has largely concerned itself with the preparation and initiation of scientists, with some concessions to the need to produce workers suited to technological occupations. It is often asserted that school science should be made relevant. Science *is* relevant. Its relevance is in its potential to fulfil very important needs for each and every one of us. The mistake is to think that a description of the dynamo, the manufacture of a fertilizer or the cross-breeding of cattle is sufficient in itself to make it relevant. The relevance of science is not necessarily self-evident (Jungwirth, 1980). Using applications of science to illustrate scientific principles may be a good teaching method but it does little to achieve the wider aims of science education. Science is relevant because it relates to people's lives. To make science *education* relevant we must point clearly to that relationship.

Science has tended to be taught in a way which makes it remote from people's lives. The result is the presentation of a body of knowledge and skills which tend to stand in isolated self-sufficiency, severed from what the students perceive to be important and valuable, either now or in the future. Does it matter? Should science education concern itself with wider aims? Does it need to be more outward-looking and justify itself more by external relevance?

Nowadays, the importance of science education for all is widely accepted. At both the primary and secondary level, the National Curriculum makes science education compulsory for all but the few with special educational needs so pressing that science lessons must have a low priority. In practice, all but a very few children of primary school age should receive some science teaching. At the secondary level, it is similarly expected that all pupils will receive some science teaching, up to 20 per cent of the timetable. Girls, traditionally shying from science, and physical science in particular, are included in that number (Omerod, 1971). Most of these pupils, boys and girls, will not become scientists or technologists. Given a choice, many may not have chosen to do science at all, so the teacher may be faced with a lack of interest and negative attitudes. Science will be taught to all, regardless of ability and inclination. It is an unavoidable part of every child's school life so it is important that it is both a worthwhile activity and that it is seen to be worthwhile. That means that the relevance of what is taught needs to be clear.

The work of the Assessment of Performance Unit has pointed to the strengths and weaknesses of contemporary science teaching and suggests the need for everday contexts to make the products and processes of science meaningful (DES, 1983). It is said that 'nothing has meaning apart from its context' (Bigge, 1982). While the science teacher's main concern is with the effective teaching of the products and processes associated with science, and appropriate contexts can facilitate the task, relevant science teaching has the potential to do more than lubricate the cogs of the mind.

In attempting to realize the potential of science and technology to fulfil needs, individuals and society meet ethical and moral problems. In Britain, it is considered the responsibility of all teachers to help to develop moral capacities and an understanding of what is meant by ethical behaviour in their pupils (Ashton *et al.*, 1975). Science education has, by and large, tended to sidestep this responsibility. Science is seen as being above morality; morality is a quality for those who apply science, not for those who create it. But more will apply it than create it, and even more will be on the receiving end. The dehumanizing tendency some see as inherent in science education needs to be counteracted (Goran, 1974). In solving a real problem, for example, due regard must be paid to the quality of the evidence, the range of valid solutions, their ramifications and the selection of an acceptably humane solution from amongst these.

This, of course, brings in a philosophical and epistemological aspect to science education. There is a need to recognize, for example, a fair test (or, perhaps more usefully, an *unfair* one), to question what is meant by the 'best' solution, to judge the quality of evidence and its limitations and to assess the validity of conclusions. These, and other such processes, are fundamental to life in a technological society. Commensurate with their age and ability, science education should provide opportunities for pupils and students to gain competence in such processes.

A wider base in primary school science is justifiable on pedagogic grounds alone: science for the younger pupil has more meaning if it is closely tied to significant contexts. There is no such thing as splendid isolation in education. In addition, to think and work in a systematic, scientific way is a particularly valuable capacity to develop. Wherever there are investigations to do and problems to be solved, scientific skills might be deployed to advantage, so science, the activity, might be found in almost any area of the curriculum. At the same time, images of, and attitudes to, science and technology begin to form at an early age and may become resistant to change unless the primary teacher ties science to the wider fabric of experience and shows its relevance to a broad range of situations (Newton and Newton, 1986).

Science plays such an important part in our lives that there is also a need to understand its role. That can only be achieved by looking outwards to the real world, rather than only inwards to the science. An appreciation of the nature of its role often calls for an understanding of science as a body of knowledge and as an activity. In school, the person likely to be best equipped to teach about that role is probably the science teacher. Since science is now a core area of the curriculum, science teachers must also consider more seriously the contribution it might make to the pupil's general and moral education. Jennings (1983) expresses well the need to do more with science teaching:

'In the long run, our only hope can lie in education: in a public educated about the meanings and limits of science and its use of technology: in scientists better educated to understand the relationship between science and technology on the one hand and ethics and politics on the other; in human beings who are as wise in the latter as they are clever in the former'.

References

Ashton, P M E *et al.* (1975) *The Aims of Primary Education: A Study of Teachers' Opinions* Schools Council/Macmillan, London.

Beveridge, W B (1980) *Seeds of Discovery* Heinemann, London.

Bigge, M L (1982) *Educational Philosophy for Teachers* Charles E Merrill, Columbus, OH.

Broad, W; Wade, N (1983) *Betrayers of the Truth* Century, London.

Capra, F (1982) *The Turning Point* Wildwood House, London.

Chalmers, A F (1978) *What is This Thing Called Science?* The Open University Press, Milton Keynes.

Department of Education and Science (1983 etc) *Science Reports for Teachers* Assessment of Performance Unit. Richard Gott, former Deputy Director of the science team, was kind enough to describe to me how he saw the relationship between relevance and the work of the Assessment of Performance Unit.

Goran, M (1974) *Science and Anti-Science* Ann Arbor Science, Michigan.

Hales, M (1982) *Science or Society?* Pan Books, London.

Hodson, D (1985) Philosophy of science and science education. *Studies in Science Education* **12**, 25–57.
Hodson, D (1986) The nature of scientific observation. *School Science Review* **68**, 17–29.
Jennings, A (1983) Biological education – the end of the dinosaur era? *Journal of Biological Education* **17**, 298–302.
Jungwirth, E (1980) Some biology/social science interfaces and the teaching of biology. *Journal of Biological Education* **14**, 339–344.
Lendrem, D (1985) What are scientists made of? *New Scientist* **1479** 57–58
Monod, J (1972) *Chance and Necessity* Collins, London.
Newton, D P (1986a) Tempered with humanity. *The Times Educational Supplement*, 18 April 1986, p.55.
Newton, D P (1986b) The world view of Linnæus. *Journal of Biological Education*, **20**, 175–178.
Newton, D P and Newton, L D (1986) Humanising primary science. *Education 3–13* **14**, 47–52.
Ormerod, M B (1971) The social implications factor in attitudes to science. *British Journal of Educational Psychology* **41**, 335–338.
Richardson, M and Boyle, C (1979) *What is Science?* Association for Science Education, Hatfield.
Roszak, T (1968) *The Making of a Counter-Revolution* Doubleday, New York.
Taylor, J (1973) *The Scientific Community* Oxford University Press, Oxford.
Ullrich, R A (1983) *The Robotics Primer* Prentice-Hall, Englewood Cliffs.
Walsh, W (1964) *A Human Idiom* Chatto and Windus, London.

CHAPTER 3

FRAMEWORK AND AIMS FOR HUMANIZED SCIENCE TEACHING

A framework

The previous chapter outlined the three main ways science can fulfil needs and therefore be relevant to our lives. Perhaps few children will be scientists or technologists but all will become citizens and are already, and will continue to be, deeply affected by science and technology.

Science courses tend to be unclear about 'making science relevant', teaching *about* science, teaching *through* science, showing the 'wider implications', and the need to achieve wider aims. The framework shown in Fig. 3.1 is offered as a starting point for constructing more detailed statements of aims. It should also help in compiling and assessing material intended to achieve what are often diffusely expressed ends. What science has to offer is a starting point. Attempts to take advantage of what is on offer have consequences; good, bad, of no great significance, foreseen and unforeseen. Such actions often involve problems of ethics and morality. It also needs some appreciation of the nature of science and technology, the difference between them and their strengths and weaknesses. What the framework does not show explicitly is that science has the potential to fulfil the needs of people both as *individuals* and *collectively*. This dual nature of the framework needs to be remembered when using it.

Teachers might choose to emphasize (or ignore) certain areas of the framework, according to their perceptions of the immediate and long-term needs of their students. Syllabuses for public examinations often do that for them. Those for GCSE science courses, for example, ask for a consideration of the economic, political, social and environmental factors, thereby focusing largely upon the middle column of Fig. 3.1. If the intention is to show the impact of science on the way we see our place in the world, then the last column might be the centre of interest. To dispel inappropriate images of science and scientists the first

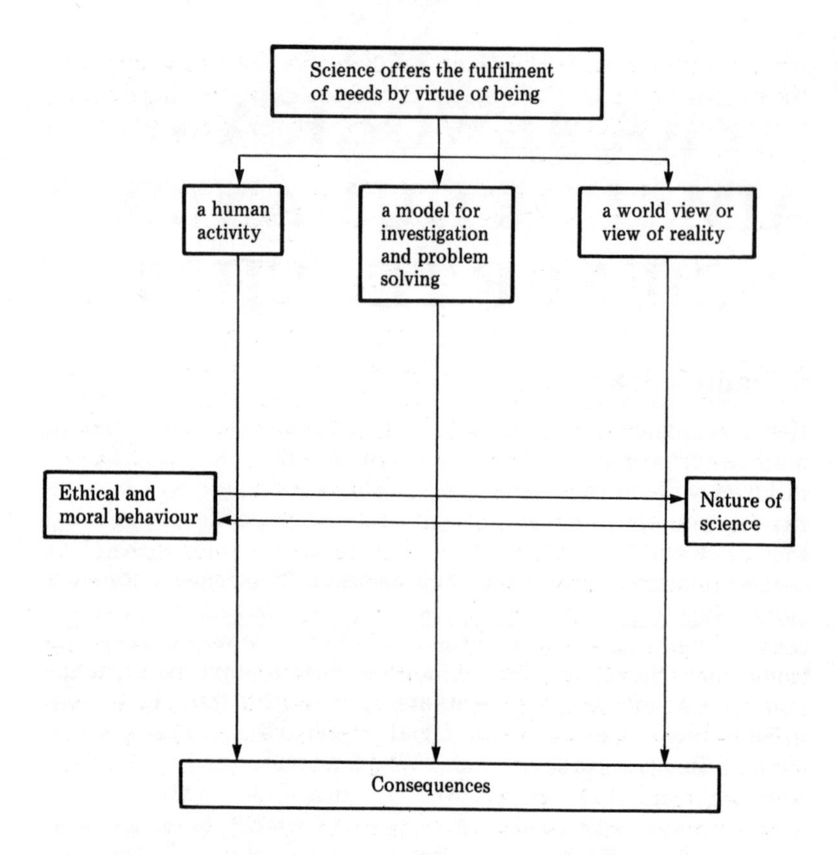

Figure 3.1 *A framework for humanized science teaching.*

column might provide worthwhile starting points.

Material written to achieve wider aims of science teaching might similarly emphasize certain areas of the framework. The *Science and Technology in Society* (SATIS) units, as their title implies, focus on what science and technology offer society, and how other considerations than scientific are involved. Since this material is intended to be used in GCSE science courses, this emphasis on the middle column of Fig. 3.1 is to be expected. What may not be expected is a variation between principal subject areas, not in the focus, but in the extent to which the material is humanized. When the proportion of clauses which refer directly to *people* is found, physics and chemistry both score much lower than biology. The count does *not* include references to Man as a biological entity. Descriptions of Man as an animal, his form and function, for example, are excluded. It seems that those who read the natural science SATIS material will be exposed to significantly more overtly humanized material than those who read the units for the physical sciences.

Percentages of humanized clauses		
Physics	Chemistry	Biology
14.8	11.7	22.9

Are the physical sciences really less relevant than biology, or is it more difficult to show their relevance explicitly? Is it, perhaps, that the relevance of physics and chemistry remains implicit in accounts of technological applications in the hope that readers will distil its quintessence themselves? This seems to be the case with the *Physics Plus* material. A similar analysis of a sample of it indicated that the proportion of clauses which refer directly to people amount to only about five per cent. The *Science Support Series*, for use in sixth form lessons, weighs in at about seven per cent, which also suggests that much of its message regarding relevance is implicit although, with older students, less teacher support may be needed. For comparison, the SISCON and Science in Society booklets, for use in sixth form General Studies lessons, gave average scores for samples of them of about 35 and 24 per cent, respectively (Newton, 1988).

Much of the more recent material, like SATIS, is mainly concerned with the relationship between science and technology and society. But the framework has a dual aspect: science may also be relevant to the needs of individuals. This often seems to receive scant attention in modern science teaching. A consideration of moral aspects, with a view to educate through science, would also be uncommon in today's science classroom. Whatever the focus, there would be opportunities for the consideration of ethical and moral behaviour across the width of the framework. The same applies to the nature of science. With younger children, these opportunities might well be in the form of direct and personal experience rather than by abstruse discourse. But such experience provides valuable foundations to build on when the time is right.

A broad approach

In Chapter 1, the four main groups of wider aims were described: moral, contextual, philosophical and epistemological, and psychological. Each seems to focus on limited features of the framework. Ethical and moral aspects might be extracted across its width to educate through science. An historical context might focus on the role of individuals while a sociological perspective might be more concerned with the behaviour of social groups. Aspects of the nature of science might be used to alter the popular image of science. Holmyard's 'anecdotal' approach, which he described as an effective generator of interest and positive attitudes (and was psychological in intent) draws on the lives of famous scientists. While, in practice, such approaches would not be straitjackets, if taken alone each could neglect important features of the framework. For example, to look only at individuals, or only at groups, would be to ignore its dual aspect.

The very significant, and often complex and subtle, roles of science and technology in our lives were outlined in Chapter 2. They affect our beliefs, attitudes and behaviour and offer us choices with the potential to affect the quality of our lives. Any person in an advanced, technological society should have the opportunity to become acquainted with that role and, if possible, should be a better person for that acquaintance. This means that, in the long run, attention should be drawn to all the major features of the framework. Each has something important to say about that role and if, in the process, they generate interest and positive attitudes – as I believe they can – then that is a most acceptable, useful and valuable bonus.

But this is not to say that, at any one time in a child's school career, all the features of the framework are appropriate. For example, it would be pointless dwelling on social issues with youngsters who have not first developed some understanding of society. The needs which science and technology might fulfil for society and the consequences of attempts to fulfil those needs would be largely meaningless. The relevance of science would not be at all apparent in such an approach. On the other hand, relevance to themselves, their families and other individuals would have meaning. As the child develops, opportunities to become acquainted with successively wider and more complex aspects of the framework would be provided. What is important is that on leaving school, at a level significant to each person, the relevance of science, its nature, and the need for rational, humane choices should be apparent to all.

On this basis, the framework is now translated into aims. While these range over all the framework, it is not suggested that they are all appropriate at all levels of education. Aims must be matched to the age, experience and ability of the learner. Some aims also appear to be hierarchical in that some acquaintance with one or more may be necessary before another can be achieved.

Wider aims of science teaching

Educating about and through science should develop:

1.0 an appropriate image of the scientist, and an awareness
1.1 that science offers a chance to fulfil personal needs,
1.2 that there is an essential commonality between the needs and sensibilities of scientists and those of non-scientists in both kind and intensity, and
1.3 that the practice of science involves the scientist in questions of a moral and ethical nature in common with many other human activities.

2.0 an awareness that science and technology are mankind's main means of adapting to the environment, and
2.1 that this makes actions possible whose intended outcome may be beneficial, detrimental, or of marginal or questionable utility,
2.2 that this makes actions possible which may have beneficial, benign, detrimental or no consequences, beyond those intended,
2.3 that, because of the complex nature of the environment and the interdependence of its parts, the choice of action must be well informed and have regard to the quality of the evidence, and
2.4 that such choices often involve values and questions of morality.

3.0 an awareness of the scientific view of reality, and
3.1 that such a view can pervade a culture and direct thought and action,
3.2 that the view is not fixed and unchanging, and
3.3 that it has inherent limitations in application which can involve questions of morality.

Each of these groups pre-supposes some understanding of what moral behaviour means. This understanding cannot be assumed to be present and it is considered to be the responsibility of all teachers to contribute to its development in their pupils (Walsh, 1964; McPhail, *et al.*, 1978). Taken together, it is a contribution to the development of 'wisdom' (Frankena, 1973).

Science, as a human activity, can be used:

4.0 to contribute to the development of moral capacities, such as

4.1 an awareness of, sensitivity to, and appropriate response to the sensibilities in the self and others,

4.2 a willingness to take another's perspective and to see something from another's point of view,

4.3 an ability to recognize situations requiring value judgements and moral behaviour, and

4.4 an ability to consider the moral acceptability of a choice and a willingness to temper rationality with humanity when making that choice.

Teaching about science will also, on occasions, touch on the nature of science (Hodson, 1985). Once again, the way in which it is treated, and the extent and depth of that treatment will be determined by the age, ability and experience of the learner.
Science education should help:

5.0 to develop an appropriate level of understanding of the distinction between scientific enquiry and other types of investigation and of the nature of science, especially regarding

5.1 the aim of science as being to describe, explain and predict observable phenomena,

5.2 the meaning of such terms as: observation, fact, law, hypothesis, theory, model, description, explanation,

5.3 scientific practice as a highly individual process of creation of concepts, hypotheses and theories, followed by a process of validation and testing by critical observation, followed, in turn, by a process of incorporation of the concepts, laws, models, and theories into the body of knowledge by the scientific community, and

5.4 limitations of science as exemplified by, for example, observation as being fallible and theory-dependent, the complexity and uncertainty of theories, the long-term temporary status of concepts and theories, without these necessarily implying unreliability and arbitrariness.

Achieving these aims – some general principles

Products, processes and people are very different species. Choose any one and make a list of what you want to achieve. Take care to put them in the order you think is appropriate for teaching. Tag on to each item what it might contribute to the aims of the other two. It is unlikely that a progression or appropriate teaching sequence will be immediately apparent for these other two aspects of science. The more fundamental aspects of the products will not necessarily coincide with the elementary aspects of the processes or conveniently allow an appropriate input of relevance. The three are essentially incommensurable. In practice, as teachers *of* science, products and processes will tend to wag the tail. When that happens, the aims relating to people will have to be achieved through the opportunities they provide. In any case, without the teaching *of* science, it is like the bark without the dog: there is nothing to make relevant.

The science teacher has a lot to do. Are these aims to become yet another burden? That would certainly be the case if schemes and syllabuses began to swell with lists of historical, social and political events, names of scientists, 'good' and 'bad' social consequences of the application of scientific knowledge, examples of altruism, fraud, scientism and anti-scientism, and so on. This would be self-defeating. It could so easily encourage a sterile and didactic exposition severed from the science content. Yet more topics to cover before the exam! Instead, teaching about and through science needs to be done on a broad front throughout a child's school life, taking opportunities to refine ideas and concepts year by year. The most appropriate way of achieving these aims seems less through a prescription of content and more through the *way* in which science is taught, being opportunist and creative in taking and making situations which allow these aims to be achieved. It is the way in which science is relevant that is more important than the material used to illustrate it.

This still leaves the question of whether the approach should be *supplementary* or *integrated*. In a supplementary approach, essentially independent material designed primarily to achieve these aims augments the teaching of science. For example, material about the Chernobyl incident could be used after an exposition on nuclear power. While the products of science, in this case nuclear energy, provided an opportunity to discuss energy problems, the omission of the discussion would leave the science exposition unaffected. In an integrated approach, it would be more difficult to tease out such material without destroying the science exposition itself. For example, fieldwork on pollution might be used as the vehicle for practising certain chemical skills and processes as well as increasing an awareness of the quality and importance of the environment.

In practice, these are extremes on a continuum. A supplementary approach often has integrative elements while an integrative one is seldom without supplementary aspects. The age, ability and experience of the learner would play a large part in deciding where the emphasis was to lie. However, I feel that in general a more integrated approach is better for younger children, since the connections are made for them and everything is tied together. Supplements can be suitable for older learners who are able to make those connections themselves and to explore the relationships with the subject matter in personally significant ways.

The dual aspect of the framework must also be considered when selecting appropriate material for a particular age-range. Young children are egocentric and their development takes them outwards, beyond the immediate family, friends and the local community to society, the nation and the world community. I suggest that the relative emphasis placed on the individual and on society should follow this development (Newton and Newton, 1987). Younger children should be offered material mainly from the line AA' in Fig. 3.2. It is pointless offering them large doses of social implications before their concept of society has been sufficiently developed. As the child develops, the progression would be to the right of Fig. 3.2, towards line BB'. Of course, this can only be a rule-of-thumb. There are complex problems of the relationship between science and the individual appropriate to high levels of scholarship while skilful teaching may render some social consequences meaningful to young children.

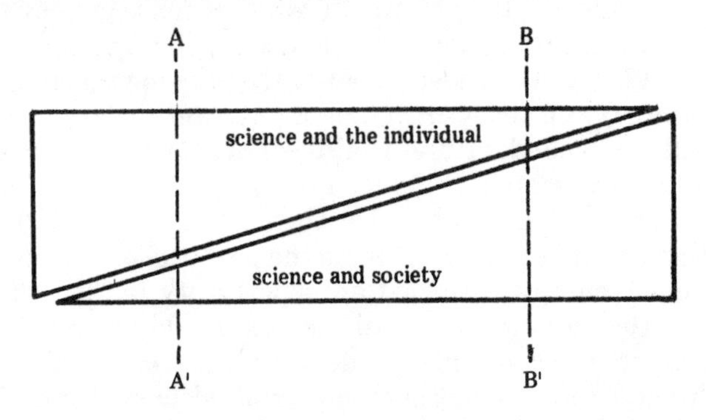

Figure 3.2 *The dual aspect of the framework related to the age of the child.*

Finally, although the relevance of science springs from its potential to fulfil the needs of people, this does not imply that this relevance must be shown with humanized science teaching. Covert and oblique approaches might achieve the same ends. For example, young children might be encouraged to care for dependent animals, group practical activities might encourage mutual consideration and cooperation, practical problem solving might provide an insight into some aspects of the nature of science and its methods. While these could provide useful foundations for later work, in the long run it is unlikely that such approaches alone would achieve as much as an overt input. Problems of morality, for example, may simply not arise in the course of a normal lesson. Social consequences may remain unnoticed, inadequate images and beliefs may go unchallenged. Oblique approaches may be useful but too much should not be expected of them. Too often, they amount to no more than the hope that something useful will be picked up. It would seem perverse not to put the people back into science and humanize science teaching when the intention is to show the relevance of science to people.

With these general principles in mind, we now turn to the achievement of these aims in the classroom.

References

Education Group Conference (1973) Social responsibility in education in physics. *Physics Education* **8**, 4–6.

Frankena, W K (1973) *Ethics* Prentice-Hall, Englewood Cliffs.

Hodson, D (1985) Philosophy of science, science and science education. *Studies in Science Education* **12**, 25–57.

McPhail, P *et al.* (1978) *Startline: Moral Education in the Middle Years* Schools Council Moral Education 8–13 Project, Longman, London.

Newton, D P (1988) Making science teaching relevant: a look at the SATIS material. *School Science Review* **69**, 826–829.

Newton, D P and Newton, L D (1987) Humanising primary science. *Education, 3–13* **14**, 47–52.

Walsh, W (1964) *A Human Idiom* Chatto and Windus, London.

CHAPTER 4

DOING MORE WITH PRIMARY AND MIDDLE SCHOOL SCIENCE EDUCATION

Introduction

Whether teaching about or through science, an integrated approach is particularly useful with younger children. The essence is to provide an opportunity to achieve one or more of the wider aims of science education at the same time that those relating to the teaching of science are being achieved. This needs no great subtlety or deviousness on the part of the teacher, merely an awareness of the wider aims and a willingness to look for opportunities.

Sometimes the opportunity does not lie in the science content or in the activities the children are doing but in the classroom organization. Teachers of younger children are usually well aware of the value of group work in providing opportunities for cooperation, mutual consideration and for respecting another's views. These are fundamental aspects of the aims of moral education (for example, relating to aims 4.1 and 4.2 in Chapter 3). Indeed, at the elementary level, shortage of materials often makes group work a necessity. It is those occasions when children *should* do practical work alone that are a luxury few can afford. However, from time to time, it is useful to try *increasing* the group size for a part of a lesson. Children learn to work acceptably with one or two others but seldom experience working in larger and more loosely structured groups. For example, it is potentially useful for children to experience working on an investigation in this way. In the initial discussion, parts of the problem might be allocated to subgroups. A typical activity might be a study of the movement of a buggy. One subgroup could investigate and determine the surfaces to be used. Another would design and construct a buggy while a third devised a fair testing procedure. All activities need inter-group consultation and the practice of skills and processes. When the group re-convenes, the units are brought together and the investigation completed. It should be added that many children do not have any great skill at working like this and the teacher may need to point the way. The camel has been described as a horse designed by a committee: children's committees can at times produce similar odd compromises. Perhaps this is a lesson in itself. If this approach is used care must be taken to ensure that some children do not miss out on essential experiences. For example, an investigation of boats might involve floating and sinking, displacement, upthrust, and effect of salinity. It might be considered that most children should experience most of these effects, so a sub-grouping approach would be inappropriate.

But when we talk of looking for opportunities, in most cases they will need to come from within the science in the lesson. The most direct way of showing the relevance of science would be to take an aim and build a lesson, or lessons, which focus exclusively upon it. In the extreme, the lesson might be independent of other work in science or, indeed, any other part of the curriculum. For example, it would be possible to use a science-technology trail like this. The children would explore the school, looking for applications of 'science', perhaps with the intention of showing our dependence on technology (aims 2.0). In practice, exercises of this kind would seldom be used in this free-standing way and would tend to be tied into other areas of study. Perhaps it would be used to show the ubiquitous utility of electricity following work on circuits (aims 2.0). This might be followed by creative writing on 'What life would be like without...' with the intention of stimulating thought about the value of some application (for example, aims 2.1, 2.2 and 2.4). The intention could then be to use this as a scene-setter for activities involving technology, for example, a challenge to invent and make an efficient and *quiet* stone sorter (aims 4.0). In this way, humanized and humanizing science teaching can be used to lead into or can come from other activities in science. This simple approach is described next. The general areas of the relevant aims will be indicated (for example, 1.0, 2.0, 3.0) where the precise aim depends on the direction of development taken by the teacher or followed by the children. More specific aims (for example, 1.2, 2.3, 4.2) will be given when the direction is likely to be more narrowly defined. It should be emphasized that opportunities to achieve others of the wider aims are probably latent in some of the examples and imaginative teachers may find those better suited to a particular class or occasion. In addition, activities which might help to achieve these aims could be developed in other areas of the curriculum. Science, as a way of thinking and working, should not be confined to the science lesson. Many of the skills and processes commonly ascribed to science are not unique to it and may be found elsewhere.

Pre- and post-science supplements

A simple and effective way of widening science teaching is to teach some science and then extend it into areas of human concern. This approach is supplementary in that, should the extension be omitted, the quality of the science taught earlier will be unaffected. In short, humanized material has been tagged on and tied into the science lesson. For example, when dealing with changes in living things with time and the match between form and function through a lesson on animal life, evolution and extinction, it is worthwhile following this with Darwin's voyage on the *Beagle*. Children are to imagine themselves to be a scientist on the *Beagle* and keep a daily diary of their experiences, observations and discoveries. Fantasies of achievement and identification with successful people are important to young children, and this exercise begins to make the scientist real and accessible (for example, aims 1.1 and 1.2).

Even when the animal life is as lowly as the worm, young children can experience a worm's eye view of the world from ground level. Essentially, this is an introduction to decentring, stepping outside oneself and seeing the world as far as is possible from another creature's point of view. (This is not only a valuable part of interpersonal behaviour, it is also a skill which can be used to suggest hypotheses in science. Older children are expected to use it in practical activities on snail colouring and camouflage. They must take a bird's eye view of the world if they are to decide which snail colours will tend to survive in backgrounds of different colour prior to counting them.) Even plant life can be treated in this way. Young children can adopt a tree, feel its texture, sit in its branches and *be* a tree. Older children might consider the significance of a notice on the grass: 'Your feet are killing me'. Is grass alive? How do we know? How can we find out? What would the world be like without grass? A respect for life and the environment is important (aims 4.1–4.3).

Closer to true role play are the scenes children might act after doing the science. For example, following a lesson on eclipses, half the class might imagine themselves as Stone Age people seeing an eclipse while the other half could present the same scene in a modern setting. Discussion before and after could help to implant the idea that the same event can elicit different responses according to beliefs and points of view (aims 3.1 and 3.2). While these might be brief events, children can, of course, act out a fuller presentation relating to their science. For instance, the science of sound might lead to some inventions of Thomas Edison. After an appropriate version of the story of his life, a simple play could be constructed which incorporates some of the science. This has the advantage of allowing the children to present or 'record' their science in an unusual, non-written form, as well as showing the technologist to be a person much like themselves (aims 1.1 and 1.2). A thumb-nail outline of such a play for upper juniors is included here to illustrate the approach. The children could write the details of the dialogue themselves. Not all biographies lend themselves to dramatic presentation and care must be taken to involve as many of the class as possible in 'scientist' roles, including girls. The lives of Marie and Pierre Curie, for example, could provide such an opportunity.

THOMAS EDISON

Thomas Edison (1847–1931) was born in Ohio. Scarlet fever in childhood left him partially deaf, which may be the cause of his lack of progress at school, and his mother had to teach him at home. At the age of ten years, he set up a laboratory in the cellar and taught himself elementary electricity and chemistry. At twelve years old he sold newspapers on a local train service and learned much about electrical devices, especially telegraphy. His scientific skills were limited and many of his inventions were perfected by trial and error but he was successful and, in 1876, he set up his invention factory at Menlo Park, New Jersey. According to Edison, genius was one per cent inspiration and 99 per cent perspiration. His most celebrated invention, the phonograph, was made in 1877, and the first recording on it was Edison's voice reciting 'Mary had a little lamb...' Henry Ford consulted Edison in 1896 to ask his opinion about the internal combustion engine as a source of power for a horseless carriage.

Dramatis Personae

Narrator	Edison's mother	Train guard
Thomas Edison	Class pupils	Inventors
Edison's teacher	Train passengers	

Narrator. This is the story of Thomas Edison who invented something found in most homes today.

Scene 1 North American elementary school in the nineteenth century. One-third of children as pupils in a class in this school. Thomas at his desk, not paying attention. Angry teacher sends for mother. Mother told to take Thomas away because he is stupid. Action which indicates that Thomas's difficulties are caused by deafness.

Narrator. Tells of Thomas's deafness, mother teaching him at home, his interest in science and his laboratory in the cellar. Introduces the next scene.

Scene 2 On a passenger train. One-third of children as passengers, seated appropriately. Thomas goes up and down the aisle selling newspapers and sweets. Guard explains to Thomas the working of an appliance on the train, for example, an electric bell.

Narrator. Tells how Thomas used his earnings to equip his laboratory and how he became good at inventing things, sells his ideas, makes a lot of money and opens his Invention Factory.

Scene 3 Invention Factory. One-third of children as inventors. Thomas supervising as they 'invent'. One inventor brings a simple phonograph to him. They try it out and discuss how it works.

Narrator. Thomas became very famous for his inventions. Some of the other things which his Invention Factory made included the light bulb, a motion film camera and an improved telephone.

Excursions into drama, for many science teachers, may be an uncommon occurrence, perhaps to be kept for a special occasion. While work on sound might take such a route, others are available and, for many, may be more attractive. The production of sound and its properties could lead to tests on sound levels using the sound-level indicator of a tape recorder, discussion of the decibel, unwanted sound and noise pollution, and the question 'When is music not music?' (aim 2.2). Within this context, there is also the opportunity to discuss the need for consideration of others (aims 4.1 and 4.2). In the same way, scientific studies of water and solutions might lead to considering dirty water, water pollution and other effects of mankind on the environment. This, in turn, might suggest more scientific work on the prevention of pollution and the cleaning of water (aims 2.1 and 2.2). Similarly, younger children being introduced to change and permanence might extend the topic to include a playground Womble. They would classify the litter, identify the most common kind and, if possible, its source. They could then carry out tests to see which was likely to be short-lasting and which likely to be long-lasting litter. This illustrates in a simple way how we can affect the quality of the environment. It could be taken a stage further by asking the children which playground they think is better (aim 2.4). Do not assume the response will be unanimously in favour of an unpolluted one. What we call rubbish, children often have a use for. Aesthetics apart, they need to see its adverse effects on plant and animal life. Visits to stinking, dead ditches are as worthwhile as visits to thriving, healthy ponds. With older children, work on air and oxygen might lead to talk of smog and air pollution to the same end.

Simple circuits with light bulbs, a popular source of activities in the science of electricity, could lead to work on 'The day the lights went out' to emphasize our dependence on this form of energy (aims 2.0). On the other hand, older children might be able to discuss sensibly 'Is the telephone a boon or a bane?' and hence see that inventions are sometimes double-edged. Perhaps they might think of other examples themselves and even some of marginal utility, like the rainbow pencil, or redundant, like the perfumed eraser (aim 2.1). The point of the exercise is that inventions involve value judgements which *they* must make (aim 2.4).

These examples are intended to illustrate how science can be extended to achieve wider aims. It will be apparent at this stage that what have been described as extensions could often be used to introduce the science. Children's attention can be captured, shaped and focused by beginning with humanized material, that is, by using *pre-science supplements*.

What did doctors do before the invention of the thermometer? The Santorio–Galileo air thermometer and its uncomfortable application as a clinical device could be used to introduce simple thermometry and tie it in to a familiar application (aim 2.1). Perhaps putting more emphasis on process skills would be an approach which could begin with a talk about magnetism, the compass and navigation (aim 2.1) in William Gilbert's time, when it was believed that garlic could weaken a magnet. The children are asked their opinion about this supposed property of garlic, leading them to devise and execute a fair test with a magnet, clove of garlic and some iron objects (aims 5.0). Ill-founded beliefs like this are excellent sources of scientific activities. As a child, I remember an old man who insisted that beetroot seeds germinated faster after soaking them in tea rather than water. How gratified I was to demonstrate this to be false and how chagrined at the old man's displeasure. Here were two lessons in one!

Starting points may lie in other areas of the curriculum. Language, for example, might lead to a poem or story about an animal, which in turn leads to drawing and writing about the live (and free) animal itself before moving on to some aspect of animal studies. Here, the precursors emphasize the property of life and the shared needs of ourselves and other animals (aims 4.1 and 4.2) before the children study them in more controlled and scientific ways. For example, a story about Hush the barn owl might be followed by observations and sketches of birds attracted to crumbs in the playground. These general observations would then be focused on the things which birds do that we also do. The children might take it a stage further and investigate (with fair tests, of course) whether or not birds have favourite foods. Eventually, bird study could come the full circle with a collection of feathers and a study of their function to find out why the barn owl in the story was called Hush. There is a poem by Penelope Farmer which begins:

> Why draw live frogs –
> They're safer dead,
> On the page
> And in our heads.

It cleverly encapsulates the tendency of biologists to make their subject the study of the dead rather than of the living (Farmer, 1975). Thankfully, there are signs that this tendency is probably decreasing in schools.

History is another fruitful source of humanized starting points. For instance, descriptions of social conditions in the eighteenth and nineteenth centuries could lead to ameliorators of those conditions. One of these, Count Rumford, created a nourishing soup for the poor:

1 oz pearl barley	$^1/_4$ oz bread
1 oz peas	$^1/_4$ oz salt
3 oz potatoes	$^1/_2$ oz vinegar
14 oz water	(1 oz = 28 g)

'Water and pearl barley boiled then peas added and boil for two hours. Potatoes added after peeling and boiled for one hour more. Stir frequently. Add vinegar and salt. Finally cuttings of bread – dry and hard – added at the last minute.'

Children might like to try a version and say what they think of it. Rumford also invented special pans and recommended improved stove and chimney designs. This readily develops into investigations of the conduction, convection and radiation of heat and everyday uses of these principles (aims 1.0, 2.0 and 4.0). Benjamin Thompson (Count Rumford) is better known for his observation of a relationship between mechanical energy and heat, noticed when feeling the heat produced by the boring of cannon barrels.

While science can grow from some other area of the curriculum, it can also move through humanized material into other curriculum areas. An example of this for the topic of electricity is given in Fig. 4.1 (aims 1.0 and 2.0; Newton and Newton, 1987a).

Integration with the science

Although valuable, supplementary approaches can, at times, seem artificial and contrived. Of course, good teachers would smooth over the joins, but there are also ways of enriching the science lesson from within. One way is to blend people and science by drawing on biographical material. Opportunities to achieve wider aims are often prolific enough for the teacher to pick, choose and seize upon a sign of interest and develop it. At times, children might not respond to supplementary material in the way expected but such an integrated approach is flexible enough to allow occasions when the bait is not taken to pass, knowing that more of the same kind lies ahead in future work. To illustrate, I have selected an extract on the life of Leonardo da Vinci, used with upper juniors. It helps to make the point that scientists and engineers are real people and that a scientific mind and creativity are not mutually exclusive (aims 1.1 and 1.2). It could also illustrate that invention is a more complex activity than just having an idea: it must be brought to fruition or else it remains an idea, as with Leonardo's bicycle (aims 2.1 and 5.0). These points would be brought out in discussion while using the material. Pictures of Leonardo's paintings might be available as stimuli, history books might be consulted to find the flavour of fifteenth century life, a child's bicycle could be compared with a picture of an early hobby horse and both compared with Leonardo's design. Discussions about how it might have worked could lead to cogs and gears and their applications. From this introduction, it is an easy step to investigate structures and forces (exemplified by Leonardo's bridges), cogs, gears and wheels (exemplified by Leonardo's

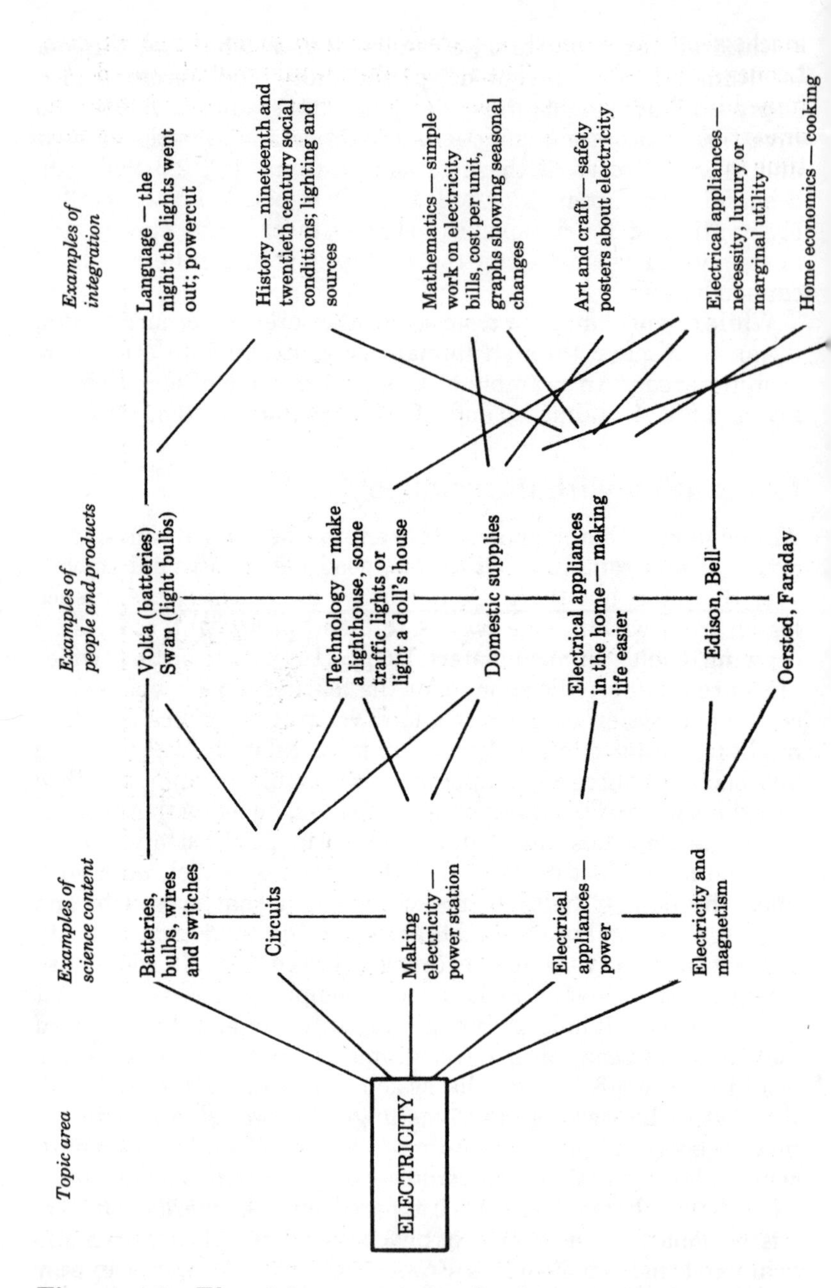

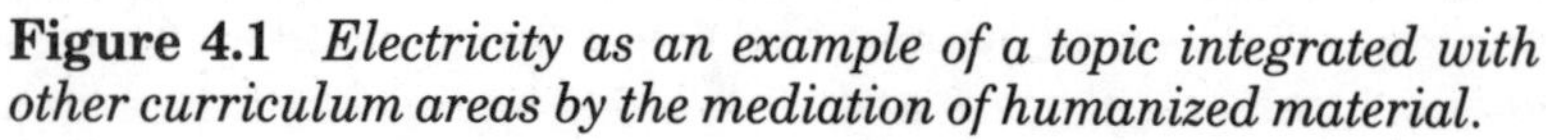
Figure 4.1 *Electricity as an example of a topic integrated with other curriculum areas by the mediation of humanized material.*

mechanical inventions) or parachutes and flight (illustrated in Leonardo's notebooks). The route taken would probably be determined by the topic under current development, either in science or across the curriculum. This might also suggest the appropriate time to use such material.

LEONARDO DA VINCI

Figure 4.2 *Leonardo da Vinci (1452–1519).*

Most people think of Leonardo only as a famous artist. He was also a scientist, an engineer and an inventor. Everything seemed to interest him.

Leonardo da Vinci was born in Italy in 1452. When he was fourteen years old he began to train to be an artist. In those days even a good artist could not always make enough money to live on. To earn a living, Leonardo had to work as an engineer for the dukes who ruled Italy. Amongst other things, he had to make plans for forts, canals and bridges. For the army, he had to invent new kinds of weapons and machines. He even thought of submarines, tanks and parachutes.

His ideas were so far ahead of their time that not all of them could be made! The picture shows his idea for a bicycle. See how much it looks like a modern one. He thought of it over 500 years ago, but it was probably never made.

Leonardo liked to be alone; that was when he had many of his ideas. At other times, he liked to be with his friends. His notebooks tell us how he sang, made up poems and played jokes to amuse them.

After he died, most of his drawings were lost or forgotten. Many of his ideas had to be thought of again by other people.

Figure 4.3 *Leonardo da Vinci's design for a bicycle.*

(Newton and Newton, 1987b)

If we want to teach about the properties of materials, we might begin with an account of the life of William Perkins and his work on dyes and perfumes. On the one hand, it leads readily to practical activities on dying agents such as lichen and beetroot, to reversible and irreversible changes, and natural and man-made dyes. On the other, it also leads to useful work on perfumes. Children might be given a selection of perfumes or perfumed soaps to carry out a class survey of preferences. Then they could see how this related to the price of the perfume or soap. Is the best always the most expensive? What do they mean by *best*? (aim 2.4). What other things do we want a soap to do? How might we test the soaps for these properties? (aim 5.1). These are a few of the questions we might ask.

The story of Roger Bacon could be used to introduce work on light: reflectance of surfaces and mirrors, for example. But Bacon also made predictions about what life would be like 700 years hence, that is, today. He predicted flying machines and ships without sails, for example. Asking the children to do the same for life 700 years from now gives an interesting insight into their view of technology as the prime agent of change in our lives (aims 2.0). The quality of the life they envisage is another matter! (aim 2.4). In addition, biographies such as these often provide children with opportunities for emulation and fantasies of achievement. Sometimes the activities arising from them offer role-play situations. The message is that science and technology are not activities only for super-humans. They are activities, even potential careers, for the children themselves (aims 1.0).

There are other, equally rich, ways of developing integrated, humanized science lessons. For example, inventions might be used. Children often find it surprising that simple, everyday objects, like the pen, pencil and pencil sharpener are inventions. They seldom think of things like these as products of a person's imagination (aims 1.1 and 1.2). The Victorian era is a rich source of amusing, impractical flights of the imagination which greatly interest children as they try to puzzle out their purpose. What we forget is that we are surrounded by the successes of the nineteenth century inventors; the camera, the telephone, flushing lavatories, the electric iron, the electric lamp, the sewing machine and the horseless carriage, to name but a few. Today's inventions are just the same. Many will fall by the wayside while a few will pass into everyday life. The Technology pages of *New Scientist* could provide modern examples to make the same point (aim 2.1).

Seizing on the interest, the teacher might give it direction with activities relating to the simple machines used as puzzle inventions, like the lever, the pulley, the incline, wedge and screw, cogs and gears. Returning to the 'puzzle', the children might be successful now and even be able to suggest why some were failures. Would *they* have used it? If not, why not? Was it superseded by something better? What made it better? (aim 2.4). Why not invent something themselves? Perhaps make something for someone who is old and cannot bend over easily (aims 4.1 and 4.2). Invent and make something out of simple things which would help an old person pick up a glove, a spoon, or a coin (each being progressively more difficult). The children might then examine the interior of a mechanical clock (aim 2.1) and relate it to their activities on simple machines. This could lead to the pendulum clock and Galileo's observation of the period of swinging chandeliers, which they might investigate themselves with lumps of clay on string. Or it could go elsewhere, perhaps to an investigation of rollers reducing friction which they had seen in their history books, or to the machines that parents use at home, for example.

Elliott (1978) suggests that for the very young, the environment could provide starting points for all the science experience a child needs. But he also warns that the environment poses so many questions that, unless care is taken, 'the shutters come down and no questions are pursued'. The particular value of the child's environment for providing an integrated approach to humanized science teaching is that the questions it poses are about phenomena and events in a total context. Whatever is investigated is first seen *in situ*. The importance of the whole for the part is more apparent.

To illustrate how the environment can provide an integrated approach to science education, consider first a rural setting. A long-disused railway line or wagonway can be a good, safe place to study flora with a view to classifying and quantifying it and comparing it with samples from the flanking fields. It gives practice in sampling techniques, such as the use of the quadrat and line transect, and in taking soil and root profiles. These give data for testing ideas about niches and re-colonization. Instead of finding the site to be a barren wasteland, the children are likely to show that it has a richer flora than that of the surrounding monocultures. We might ask if they think this is a good thing (aims 2.1–2.4). A search for signs of fauna might show that the route is now used by wildlife. The children may appreciate the irony that what was once constructed as a social and commercial artery now serves as a footpath for its first occupants. Furthermore, its value to them is probably all the greater as much of the hedgerows and woodland have been uprooted since the line was built (aim 2.2). We might also ask how these disused lines might be reclaimed, if it is feasible and if it is desirable (aims 2.1–2.4, 4.4).

In an urban setting, we could use a street study to collect information about the uses and properties of materials and the strength of structures. Consider only a limited aspect of this in which the materials and construction of buildings are examined. Children would collect information about the bonding pattern of bricks and the size of supports in different materials, for example. At the same time, they would observe the crumbling of stone effigies in the exterior niches of a church. Are those inside crumbling, too? They would see the blackened stonework. Is it only skin deep? Most of these observations pose the question: Why? (aim 2.2). And there would be the graffitti. Back in the classroom, the children would test the strengths of different bonding styles with thin strips of wood and a safe adhesive. They could make miniature concrete beams for testing under compression, extension and shear. Their conclusions would be related to their observations. Have they answered their questions? Do they need to carry out further tests? (aim 5.1). What about those graffitti? Was paint invented for that purpose? (aim 2.2). If they thought graffitti were a disfigurement of their surroundings, they might try to make a graffitti-proof brick. The brick would be treated in different ways, perhaps rubbed with a candle, shoe polish and soap, to make a fair test to see if any of the treatments was effective. Would the treatment be feasible or acceptable on a large scale? (aims 2.1, 2.3 and 5.3).

Even those schools in seemingly dreary surroundings can often turn them to advantage. A third year junior class in a primary school near a coke works was engaged on a project on pollution. The effluent pipe from the works discharged warm water into a stream so the class tested the water for temperature above and below the pipe. They found the water below to be several degrees warmer than that above. It was also more acidic (for these children, it was more 'sour'). The flora and fauna in the water and on the banksides were sampled above and below the outlet. Marked differences were noticed. Life was varied and seemed to thrive above the pipe. Below, it was limited and much less prolific. Other causes were considered. The run-off from a nearby waste heap was a possibility. It was tested, too. Tentative conclusions were drawn, opinions expressed on the moral and aesthetic aspects and the class returned to school to pursue their investigations.

The environment in most parts of Britain has been modified and is still being modified by Man. These examples illustrate how this environment is rich in science activities and opportunities for achieving wider aims. Often, these are so inter-related that the environment provides a truly integrated approach to science teaching.

Planning to achieve wider aims

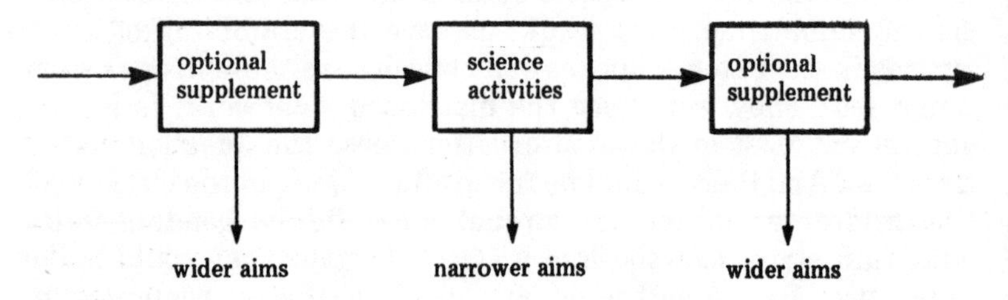

Figure 4.4 *A linear, supplementary approach.*

At its simplest, the supplementary approach is a linear one (Fig. 4.4). The integrated approach, on the other hand, is inherently more complex. At its purest, the teaching of, about and through science would be achieved through the same piece of material (Fig. 4.4). This also makes it an economical approach to use since it offers opportunities for achieving a multiplicity of aims in each teaching unit. It is probably a more efficient way of teaching children since the various facets of the science all hang together on the same backcloth. To set against that is the time it takes to construct such a unit, unless published material is used. A biography, for example, might show a scientist as a human being and, at the same time, provide an understanding of some scientific generalization, describe an instance of its practical utility and expect the child to make a simple, reasoned value judgement. The practical activities which stem from it would reinforce that understanding, practise skills and processes and help towards providing a feeling for the nature of science. The activities could be given a purpose intended to enhance the child's moral capacities.

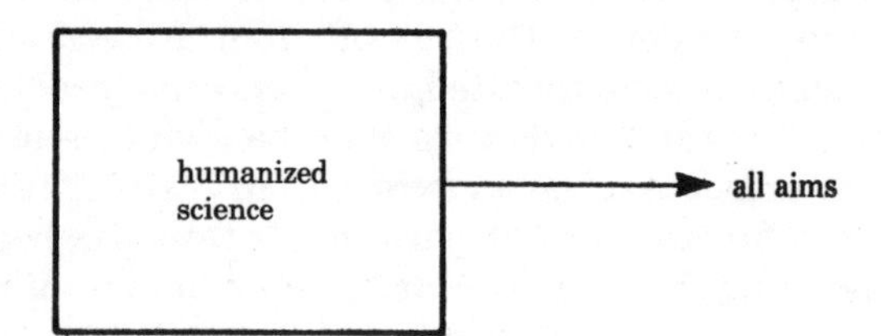

Figure 4.5 *The integrated approach.*

In practice, the distinction between the two approaches would tend to be less clear cut. Supplementary approaches which stimulate the children into making their own contributions and suggestions for scientific investigations are particularly valuable. Most teachers would try to use these opportunities so that the linear approach would become branched and lead away, at least for a while, from the proposed route. When this happens, the approach can develop into a more or less integrated one. At the same time, an integrated approach is seldom without its supplementary aspects which, by definition, could be removed without greatly affecting the teaching of science.

Elementary science teaching lays important foundations of knowledge, understanding, skills and attitudes. I believe that these foundations will be flawed if the children do not become increasingly aware of what makes science relevant. At this level, the wider aims are largely to do with relating science to the child and the groups of which that child is a member. The scientist should be seen as a sentient being with emotions and feelings that the child can recognize and sympathize with. Opportunities to show the role of technology, especially in the child's own life, and its potential for change should also be taken. Value judgements in choosing between the alternatives science and technology might offer should also be practised in simple ways. Not all aspects of the aims are necessarily appropriate at this stage; it would be pointless, for example, to discuss world views before the children have reached a stage of development and level of experience where this would be meaningful. For the same reason, abstract discussions on the nature of science would be inappropriate. Instead, the need to confront ideas with controlled experience, the need for fair tests and an awareness of the limitations of generalizations should be appreciated. In this respect, open-ended investigations are often useful devices for developing that appreciation and illustrating the complexity of real-life situations. The social implications of science are also not appropriate until an adequate concept of society is well-established. This is why some aims have received little emphasis in the examples given earlier. It is also important to place more emphasis on those aims which underpin others. For example, it would be worthless to ask a child to consider the moral acceptability of a choice if he did not recognize it to be a situation requiring moral behaviour.

What is important is that opportunities should be taken to humanize elementary science teaching. These opportunities should, as far as is appropriate, be spread across the range of aims and should concentrate on developing an awareness of the fundamental aspects of those aims. This is especially important when they are hierarchical in nature. As well as being complete in itself, elementary science also underpins the science that follows.

Since it is likely that the organizing principles and basis of progression in science teaching will be based on the products and processes of science, rather than on people, it is easy to imagine that provisions for wider aims are adequate. It is useful to have a checklist of such aims and note the occasions over a given period when each was aired. However good our intentions, it is easy to develop an imbalance and care must be taken to ensure an adequate representation during a school year.

However well these foundations are laid, they would crumble and dissolve unless maintained and developed by the science teachers of older children. This is the theme of the next chapter.

References

Elliott, C (1978) Science. In Carson, S McB (ed.) *Environmental Education* Edward Arnold, London.

Farmer, P (1975) Poem. *Journal of Biological Education* **9**, 4.

Newton, D P and Newton, L D (1987a) Humanising primary science. *Education, 3–13* **14**, 47–52.

Newton, D P and Newton, L D (1987b) *Footsteps into Science* Stanley Thornes, Cheltenham. This is a humanized science scheme for the primary school with children's workcards and a teacher's guide.

CHAPTER 5

DOING MORE WITH SECONDARY SCHOOL SCIENCE

Introduction

In the secondary school, examination constraints and the expectations of society, parents, governors and even of headteachers, can lead the teacher to adopt an overly narrow view of what science education is about. The magnitude of these pressures probably gives a secondary science teacher less freedom than a colleague in a primary school. Any concern for wider aims in such circumstances is likely to be an early casualty in the push for examination success. Consequently, methods used to educate in science must have, at least some of the time, by-products which widen that education and which educate. Or, educating *about* and *through* science must usually contribute significantly to education *in* science. An inevitable and realistic emphasis on the teaching *of* science is to be blended with an honest attempt to achieve wider aims.

There is nothing described in the approaches for teaching science to younger children which confines their essence to the primary phase. After an appropriate allowance is made for stage of development and level of experience, such approaches can often consolidate what has been gained, provide a continuity of approach at the time of transition from one level to another, or can be used to achieve wider aims at successively higher levels. For example, it was suggested that young children might attempt to act out episodes in the lives of scientists. Professor Eakin, at Berkeley, makes the 'great scientists speak again'. He takes on the roles of scientists like Mendel, Pasteur and Darwin, speaks their words and acts out significant periods of their lives. His students re-live 'the excitement of their discoveries' (aims 1.1 and 1.2) and pursue the subjects revealed in later lectures (Eakin, 1975). The point is that the vehicle is the same although the level is different. In both cases, the intention is to bring the scientist and his work alive for the learner. What is achievable depends, among other things, on the particular scientist portrayed. As well as aspects of his life and personality (aim 1), contemporary social influences and prevailing world views may be relevant (aim 3). For example, Linnæus' classification system was built on a fundamental belief that God had created basic natural units from which all variety springs (Newton, 1986). If Eakin's play included the creation, validation and incorporation of the generalizations into the body of knowledge, then it should be possible to point to these phases and to highlight particular details within them (aims 5.0).

Many of the approaches described earlier, like this one, are still useful at this level and variations of them will be referred to later. With younger or less able secondary school children, they can be used with little or no modification. As in the previous chapter, the general areas of the relevant aims (for example, 1.0, 2.0, 3.0) will be given where the precise aim depends on the direction taken. More specific aims (for example, 1.2, 2.3, 4.2) will be indicated when the direction is likely to be more narrowly defined.

Supplementing the science

It will be remembered that the easiest way of making provision for the wider aims of science teaching is to supplement the science with humanized and humanizing material (Fig. 4.3). For example, following elementary work on machines and on pulleys in particular, children might investigate the wheelless Viking pulley block (Fig. 5.1). How could it be used? What is its efficiency? Is it worth using? In what ways, if any, would a wheeled pulley block be better? The general thrust is towards loosening the popular belief in a simple relationship between science and technology. The Viking pulley block serves its purpose. Compare its action with the popular school science activity using a rope and two broom shanks. How much 'theory' went into its design is only conjecture but it could have been based on just such a simple observation (aims 5.0). The value of science in this case is that it can predict what will happen when the design is varied (aim 5.1).

With older students, the direction might have been different. The efficiency and power of machines and aspects of their automatic control might have led to an account of Descartes' 'robot' daughter and Vaucanson's duck. There was the belief that if the functions of an animal, a duck say, were adequately provided for

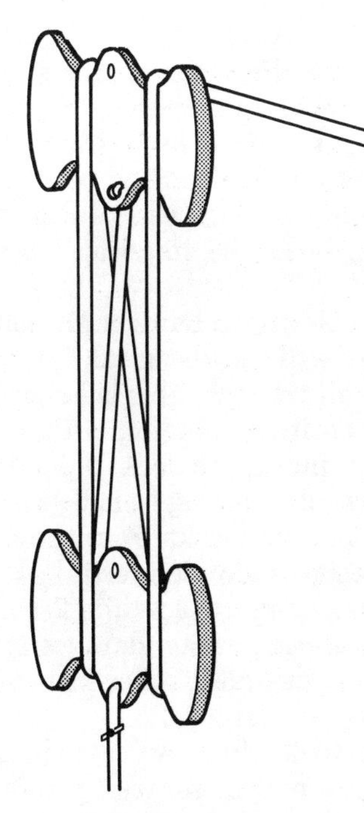

Figure 5.1 *The wheelless Viking pulley block.*

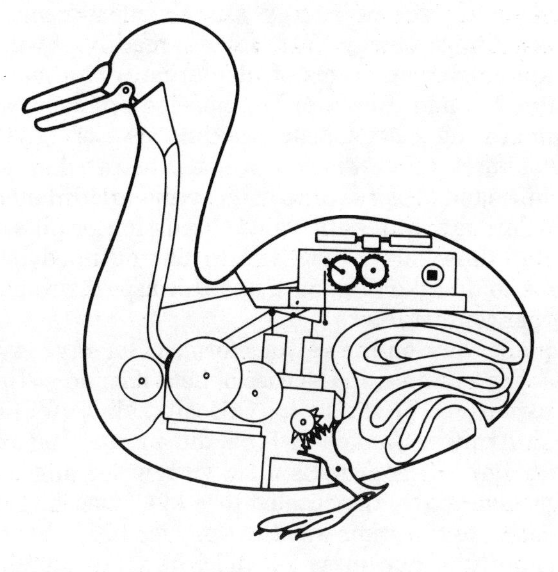

Figure 5.2 *Vaucanson's duck (after Strandh, 1982).*

mechanically, then that animal would live. Vaucanson's duck was such an attempt; it could eat, move and excrete (Strandh, 1982). A mechanistic view of nature, given form by Descartes and used so effectively by generations of later scientists, permeates the Western world view. We might also ask: suppose Vaucanson's duck had lived? What might have been the implications and ramifications for mankind? (aims 3.1–3.3).

During a study of momentum and Newton's Laws of Motion, it is very likely that colliding bodies will be discussed, possibly illustrated by collisions between small trolleys. This is helped by reference to a driver's tendency to continue in motion after a car crash and to seat belts which use inertia to lock them in a collision (aim 2.1). An additional hazard resulting from this is the risk of fire from the ignition of petrol by sparks from the car's electrical system. Students could design a simple inertial switch which will turn off the electrical system on impact (aim 2.1). We might ask why are they not fitted to all cars as standard items? It illustrates that although a solution is available other factors (in this case, financial) enter into its adoption (aim 2.4).

The topic of work and energy benefits from an introduction based on a wide-ranging discussion of energy sources, resources and their limits. As well as relating the topic to life (aims 2.0), it helps to build a feeling for the concepts involved before the study is focused on classes of energy and its measurement. House design to minimize energy waste follows readily. It will probably concentrate on various forms of insulation with their beneficial effects (aim 2.1) and it is useful to mention problems which may arise from sealing a house, such as condensation (aim 2.2). The principle of energy conservation can be illustrated in many ways but should include reference to solar, wind and tidal energy, for example. This facilitates informed discussion of alternatives to fossil fuels (aims 2.0). Elliott (1978) has outlined routes such discussions could take (Fig. 5.3), and this aspect will be returned to in another context later.

Subsequent work on the various forms of energy may be about its effects. For example, in the case of heat it could be the study of thermal expansion. To satisfy his curiosity, Major Williams, with the British Army in Canada in 1785, did a gratifying experiment with two hollow cannon shells filled with water and sealed with iron plugs. Overnight, they cooled to −28°C. One shell was burst while the plug of the other was thrown over 100 m by the force of expansion of the water (aims 1.1 and 1.2). The bimetallic strip is used in a number of control devices. One thirteen year old

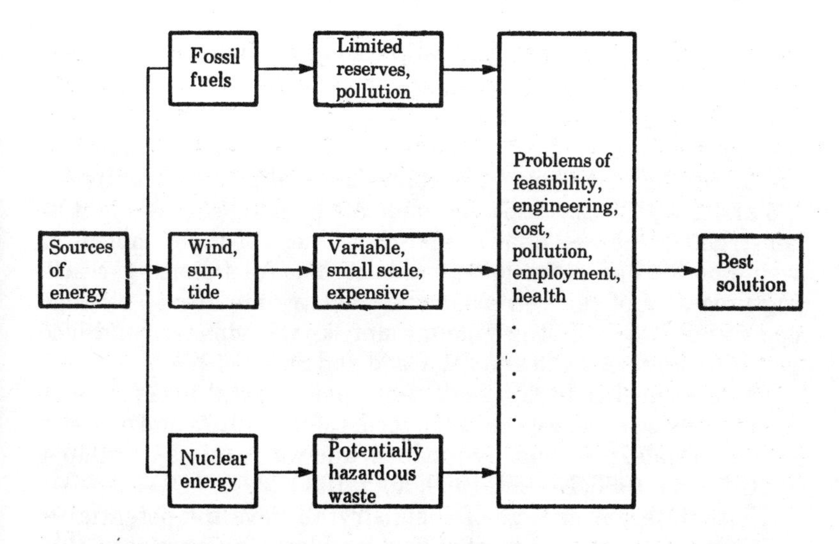

Figure 5.3 *Elliott's flow chart for the topic of Energy Sources (Elliott, 1978).*

invented a bimetallic device for indicating the depth of tea in a cup for a blind person (aims 2.1, 4.1 and 4.2). We might bring the pupils' attention to the thermometric strips used to indicate a child's temperature by pressing it to the forehead. Why are these more appropriate than the usual thermometer? (aim 2.1).

These are examples of how physics is extended or supplemented to provide opportunities for achieving wider aims. The supplementary material can often serve as a starting point, too. A newspaper article about an earthquake could introduce the need to detect and predict such events (aims 2.0). It would conveniently start a lesson on inertia and pose the problem of how to use inertia to detect earth tremors (Smith, 1986). In this case the relevance of physics to large numbers of people is being made very real and explicit.

In chemistry, the supplementary material might be water pollution when dealing with water. To be precise, it could be an account of what has happened to Lake Michigan (aims 2.2, 2.3 and 2.4; Collinge, 1972; Burton, 1979). In turn, it could lead to sampling in local streams and a qualitative analysis of the water to detect dissolved salts. The fieldwork might include observations of the action of the water on the stones in the stream bed, on precipitates and scum. Follow-up work could try to identify their causes. Are they natural or man-made? Are they benign or poisonous? Does the stream contain wildlife? Did it do so within living memory? These are some of the questions which might be asked with the requirement that answers will be reasoned. Similarly, oxidation, combustion and their products might lead to air pollution, the Greenhouse Effect and thermal pollution (aims 2.2, 2.3 and 2.4; Pearce, 1986). A simple air-filtering device might be constructed and the quality and cleanliness of the sampled air would be the basis of follow-up science. The fractionating, cracking processes of the petrochemical industry might also offer an opportunity to talk of petroleum, anti-knock additives and lead pollution (aims 2.2, 2.3 and 2.4; Gadd and Smith, 1977).

The chemical industry itself can provide useful ideas for supplementing science, especially in terms of the nature and origin of raw materials, products, by-products and waste (aims 2.0; Bates, 1980; Lewis, 1980; Nellist, 1980; Reid, 1980).

These different aspects of chemistry all have the potential to lead towards one of the most serious problems facing society; that of pollution. In some cases, the technological side of the problem has been solved but other factors are involved. Large problems are usually large because they are complex (aims 2.3, 2.4, 3.3, and 4.4). As far as teaching method is concerned, this extension material could equally have been used as a starting point, a relevant context from which the chemistry developed. For example, a report of a house being destroyed by an explosion of methane from a waste tip (aim 2.3, Fricker, 1986) could lead to the chemistry of methane and what should be done when it leaks from such sites. Would burning it add to the Greenhouse Effect? It serves to point to the possible complexity of real problems and the difficulty of predicting the ramifications of a proposed form of action (aim 2.3).

Holmyard's anecdotes were a form of supplementary material. His approach can provide variety and apparent spontaneity. For example, we might mention the coining of the word gas (chaos, from the Greek χαoς) by Van Helmont and ask why it is so appropriate. The clinical precision of a scientific term can conceal the hand of an imaginative creator (aims 1.1 and 1.2). Similarly, in connection with soap or potassium, we might mention the use of wood ash as a source of a washing agent. Is this an instance of scientific understanding *following* a useful practice? Once science has given us that understanding of what is happening, what role might it have next (aims 2.0 and 5.0)? In the same way, we might mention the use of phosphorus in early matches and the 'phossy-jaw' consequences of biting them. The 'conversation' might then turn to the safety match and its invention by John Walker in 1826. His 'friction lights' were a mixture of antimony sulphide, potassium chlorate, gum and sulphur, and sold at fifty for a shilling (aims 1.0, 2.1 and 2.2).

An excellent, modern version of Holmyard's approach is John Lenihan's *Science in Action*, (1979). He captures its essence when he writes in the preface; 'In today's frantic pursuit of innovation, prosperity and power it is easy to forget that the practice and study of science can be entertaining – even amusing – and that the pleasure can be shared without deep understanding or close involvement' (aims 1.0). In relation to chemistry, Reid (1980) has also described some of the 'chemical interfaces' with people (Fig. 5.4). It is a useful diagram to refer to when collecting supplements for a topic.

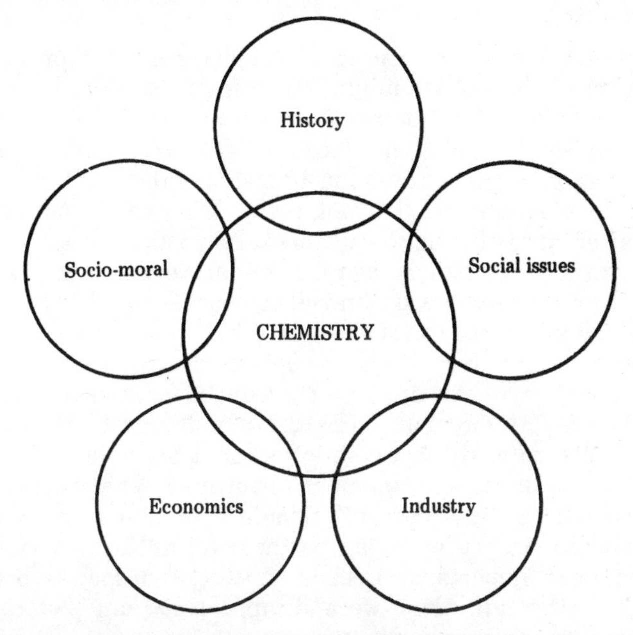

Figure 5.4 *Reid's chemical interfaces (Reid, 1980).*

In biology, an emphasis on function helps to highlight the shared needs of life forms as well as their diversity. The nutritional habits of the panda and of the nineteenth century Irish have common qualities. The failure of the potato crop was a disaster for the Irish, it being their staple food. But they could adapt. The staple food of the panda is bamboo, which is declining in availability. So, what of the panda? Can it adapt? Is it sufficiently like us that it can readily change its habits? Should we try to do something about it? What is the cause of the decline? Often, the answer to the last question has to be that, directly or indirectly, it is man who is to blame (aims 2.3 and 2.4). There are many examples we might cite: the extinction of the Passenger pigeon, the Tasmanian tiger and the dodo through hunting, are well known (Douglas, 1986). The world view which accepted this action is exemplified by the account in *Nature* on the 3rd February, 1870:

> It may interest some of your ornithological readers to know that a specimen of White's Thrush... was shot near here... and presented to Mr. Cecil Smith, of Lydeard House, for his collection. This is, I believe, the fourth occurrence of this bird in Britain; it is, I think even less known on the Continent.

Such examples could be used to supplement work on ecology, niches, or evolution. We might ask to what extent are our views the same today? Local nature and 'non-nature' trails might suggest an answer which is not too simplistic (Wray, 1968). Nature trails exemplify good intentions, but what of the problems caused simply by too many people using them? What will be the effect of class after class of biology students taking samples? Good intentions can go unwittingly awry. Non-nature trails to show, for example, untreated sewage and oil on beaches could be lessons on choice and values (aims 2.4 and 4.0).

The general aim of such supplementary material is to stimulate thought and discussion about the function and value of our shared environment (aims 2.0 and 3.0). Children should develop the capacity to consider, without sentimentality, such questions as: Is the environment important? What is it that we value about the environment? Should it be managed/protected/conserved? What do we mean by these terms? Is variety in the environment important? Does it matter if a species becomes extinct? After all, they were disappearing long before man became a potent force in the environment.

Looked at as a resource, diversity may be justified on utilitarian grounds. It is a source of genes for developing new strains of plants and animals resistant to pests and diseases, it provides us with important pharmaceutical products from plants, animals, fungi and micro-organisms, it offers us structural materials, raw materials for industrial chemistry and for fuel. Here, conservation is seen as a form of natural thrift and the husbanding of potential resources (Wray, 1976). This argument is readily appreciated by younger students but, of course, it is a limited one. The real justification for something does not lie only, or even necessarily, in its utility, but in the needs it fulfils. At some point, *education* demands that the child comes to terms with this. Ehrenfeld (1986) lists some of the reasons given for preserving biological diversity: religious, ethical, cultural, sentimental-historical, intellectual and aesthetic. Taken together with the depletion of mineral resources, this amounts to a need for the conservation of physical, biotic and psychological resources. Children need to see value in things beyond that of utility. The environment offers a good lesson in that (aims 4.0).

At a more abstract level is the view that life is a complex whole of inter-related and inter-dependent units in a state of natural balance. On this basis, the greatest care and thought is necessary before this balance is disturbed. Some theorists take this a stage further, to include both animate and inanimate in the maintenance of that balance. They compare the Earth as a living entity with the American redwood tree, which is 90 per cent dead. This *Gaia* concept, largely due to Lovelock, could be fruitfully explored by older students by reading and discussing his short account of the theory (aims 5.0; Lovelock, 1986). They might speculate on the new balance which might arise if a species (for example, man), or even a kingdom (for example, plants), ceased to exist (aims 2.2–2.4).

Not all supplementary material will be so ambitious. The mechanics of locomotion might provide an opportunity for younger pupils to design and make a simple, artificial limb for a semi-paralysed person (aims 2.1, 4.1 and 4.2). Work on respiration might lead to the effects of cigarette smoke and to the morality of smoking in the presence of non-smokers (aims 4.0; Searle, 1980). A study of reproduction could lead to discussion about over-population or test-tube babies (aims 2.0 and 4.0; Dessel *et al.*, 1973), while the ubiquitous exemplar, the Anopheles mosquito and its role in the spread of malaria might be extended to include its control with DDT. While controlling these insects and, therefore, the disease, DDT moves up through food chains until its concentration is sufficient to affect the survival of the animals at the top. When this powerful insecticide is not used, malaria tends to increase. How do we resolve the dilemma? (aims 2.0).

As in the case of the physical sciences, supplementary material also often lends itself to use as a scene-setter for a science lesson. Many of the above examples could have served this function equally well.

Extending or introducing science with humanized or humanizing supplementary material is a relatively straightforward way of providing opportunities to achieve wider aims. At its simplest, the material is grafted on to the science lesson. There are a number of more comprehensive approaches which can be integrative in nature. These seem to fall into loose groups and are described next. In practice, supplementary approaches could be developed into more comprehensive ones while some of the latter might be edited for use as supplementary material.

Integration with the science

TOPICAL OR 'BIG' ISSUES

In this approach, issues and events which are of current significance are taken as the basis of the science lesson. The Chernobyl incident, for example, might become the heart of a programme of work on nuclear energy which, in turn, becomes a part of a wider exploration of energy problems, fossil fuels and renewable sources (aims 2.0; Harris and Osborne, 1978). In a sense, the incident provides the reason for what follows it and shows it to be of immediate relevance. The science which arises from it is necessary for a fuller understanding of the subject (aim 2.3; Crawley, 1975).

Air and river pollution and the polluting effects of pesticides are recurring themes of public debate. They could become the organizing themes for work in chemistry on gases, solutions and purifying techniques, and in biology on food webs (aims 2.0; Dussart, 1981). Sixth formers might develop these topics at a higher level or tackle the acceptability of risks as exemplified by Minimata disease and the Seveso incident (Burton, 1979), or the same end might be achieved by a case study of pollution by polychlorobiphenyls and the accumulation of chlorinated pesticides in wildlife and human tissues in Sweden and Britain (aims 2.2, 2.3 and 2.4; Bartle, 1976).

This approach illustrates particularly well the relevance of science to everyday life and to the complexity of the problems it is expected to solve. Social, political, industrial, economic, environmental and personal factors are present to varying degrees in such issues.

The problem is that of availability of information. The teacher must work hard to accumulate data and background, all of which soon date and seem less immediate to the student. The Torrey

Canyon oil pollution incident, for example, seems less relevant to those who are too young to remember it than something in today's paper (aims 2.2–2.4). Not all material dates so quickly. Finding out about the chemistry of Concorde is a topic which could be repeated for as long as Concorde is not superseded (aim 2.1; Jones, 1980). Similarly, the commercial biosynthesis of protein from oil in the Protein Project, thermal insulation in the Rockwool Project and the process of hydrocracking in the Grangemouth Project are likely to stand the test of time (aims 2.0; Dowdeswell, 1981; Harris, 1981; Scott, 1981).

HYPOTHETICAL CASE STUDIES

Topical issues are not in the control of the teacher. Opportunities for achieving particular aims arise – or do not – as a matter of chance. Hypothetical case studies, on the other hand, have the advantage that they can be tailored to provide the opportunities. Being hypothetical, they need not be tied to current events so closely and so do not date in the same way as topical issues. They can, of course, still relate to 'Big' issues. The Ridpest File, for example, is a case study of the use of pesticides to protect crops and public health and involves 13–16 year-olds in making decisions about their use (aims 2.0; Dowdeswell and Wells, 1977). Similarly, Mungo Island, by M H Hansell, describes the flora of a vanished island; the children are to provide its fauna (aim 2.3; Spencer, 1977).

SIMULATION EXERCISES

Just as real issues can be made to blend into hypothetical case studies, or underly them, hypothetical studies shade into games and simulation exercises. The latter are characterized by an element of chance which alters the course of the exercise as it progresses. For example, there is the Energy Game, designed for use in the Scottish Integrated Science Course for children of 11 years old upwards. It illustrates energy concepts and the use of energy in the world around us (aim 2.1; Campbell and Drummond, 1977). For sixth formers, on the other hand, there is the Amsyn Game, emphasizing skills in problem solving, decision making and the limitations of science in relation to a town dependent on the manufacture of aromatic amines (aims 2.3, 2.4, 4.3 and 4.4; Spencer, 1977).

Kempton and Allsop (1985) have pointed out the danger of simulations being too remote from the pupils' experience to be relevant. They prefer to develop such exercises from local material, hence their Oxfordshire Coal Mine Simulation Game which considers energy, pollution and the Greenhouse Effect set against the needs of the local community (aims 2.3, 2.4, 4.3 and 4.4). Their point is a valid one, especially for younger and less experienced children. When commercial simulation exercises are used then efforts must be made to relate them to the local situation.

Project work

While most of these more complex approaches have involved children in an active and varied way – in thought, discussion, writing, card and board games and in role play – they do not have to be confined to these. Practical investigations of aspects of the topic are often possible. Sometimes they provide direct experience of the subject itself, offering ways to develop the science and the wider aims simultaneously. An investigation of local river or air pollution is such an instance. Theory, experiment design, sampling and testing in real situations which can be readily extrapolated to those that are country-wide and world-wide are invaluable (aims 2.0, 5.0; Collinge, 1972, 1974). The Pimlico chemistry trail is a project on city air pollution and its effects on building materials (aims 2.3, 5.0; Borrows, 1984). Gypsum: A School-Industry Science Project (May *et al.*, 1980) relates school science to local industry (aims 2.0). Towards similar ends, projects on food dyes, soaps, toothpastes, viscosity, road salt and lead in soil and dust have been devised (Greatorex and Lister, 1980).

Older students can investigate most of these topics in more depth, with more sophisticated experimental techniques and wider knowledge and experience. Pollution investigations, for example, could involve quantitative measurements of amounts of lead, the pH of environmental fluids, the construction of carbon dioxide detectors and the quantitative effects of pollution on stone, cloth and rubber (aims 5.0).

These more complex approaches are all to do with achieving the aims of science teaching and, *at the same time,* taking opportunities to achieve wider aims. While they have been presented in discrete categories, in reality the total approach might be a blend of several. A topical issue might have embedded in it a simulation role-play game and some practical project work to aid understanding of the issues involved and to relate them to local conditions. These approaches are largely characterized by the way they draw on contemporary issues and problems. As at the primary level, highly integrated and economical approaches can also be devised which draw on historical material.

A BIOGRAPHICAL APPROACH

It is important to emphasize that material can seldom be described as purely for primary, lower secondary, upper secondary or sixth form pupils. It is the way it is used that determines the level. To illustrate; the life and work of Leonardo da Vinci was used as a vehicle for teaching primary science and technology. Presented in appropriate language, it would be equally suitable for use at the secondary level (aims 1.0). We might, of course, take opportunities to achieve different aims and we would probably take a different direction with the practical activities, especially where the sciences are differentiated into separate disciplines at this level. In physics, for example, Leonardo's investigation of the range of a bow could provide a background for work on mechanical energy (Heydenreich, 1980).

There is a tendency in secondary science to expurgate personal reference. What does Dalton's Atomic Theory mean to pupils? In most cases, it means nothing more than billiard ball-like entities 'discovered' by some faceless stranger called Dalton. In that case, why bother to ascribe the theory to Dalton at all? His name serves no other function than that of a label. But appropriate personal details help to prevent the development of inhuman stereotypes in the minds of learners. They begin to see that scientists are sentient beings (aims 1.0). Often, textbooks present their work in such a laundered form that it no longer seems to be the work of a fallible human being (aim 5.4). Even the diagrams of apparatus lose the feel of being made by people (aim 5.3). How many of us have been surprised by the difference between the actual equipment used by a scientist and its textbook illustration? Obviously, realism can be taken to the extreme where it confuses rather than illuminates, but few text writers make the effort to communicate that feeling of personal contact.

One school achieved this in a week given to learning about Humphry Davy as a bicentenary event. As well as involving the Chemistry, Physics and Biology departments, the Humanities department found that they could contribute too (Barratt and Stanyard, 1979). Each science has its 'Greats' to draw on. In elementary physics, there are Joule, Rumford, Dewar, Gilbert, Newton, Swan, Franklin and Volta, for example. For chemistry, the list might include Dalton, Priestley, Marie Curie, Lavoisier, Baekeland and Perkin. Biology could be represented by Linnæus, Cuvier, Darwin, Mendel, Pasteur, Jane Goodall, Watson and Crick. However, the aim is not to teach the history of science but to use it to achieve the aims of science teaching. Lesser scientists, therefore, might sometimes be found to provide more opportunities in this respect. Certainly, contemporary scientists should be considered both because they include more women than in earlier times – for example, Lynn Margulis (Keller, 1986), Dorothy Hodgkin (Wilkins, 1986) and Barbara McClintock (Cherfas and Conner, 1983), and because they show the importance of team work in modern science.

It has already been mentioned that Professor Eakin introduces the lives and works of biologists by acting out episodes of their lives. Older students can also gain a lot from biographical studies. The life of Linnæus, for example, exemplifies how a particular world view determines what is seen to be reasonable or even possible (aims 3.0; Newton, 1986). Physical science students may also find Ampère's life and work of interest (Gee, 1970), and that of Blondlot and his N-rays or Barkla and his J phenomenon (aims 1.0; Hecht, 1980; Wynne, 1979).

Students have a tendency to direct their thoughts inwards on a subject and to compartmentalize it, even with biographies. It may be possible to reduce this tendency by asking them to compare the work of a scientist with a non-scientist. We might ask, for example, who had more impact on the world, Faraday or Napolean? Darwin or Marx? Many such pairings are possible.

Actual case studies

While the biographical approach tends to centre upon the life and work of a given scientist, the case study usually draws its lessons from the evolution or development of a particular product of science. Particularly useful are those case studies which require the active participation of the student. Some of the *Historical Case Studies* series of *Physics Education* are like this. *Undercurrents*, for example, describes the work of Faraday, Babbage and Herschel in connection with electromagnetic induction (Newton, 1980). It includes practical activities, original data, integrated questions with clues but few answers. The following extract illustrates the approach.

BABBAGE AND HERSCHEL'S EXPERIMENTS

A fascinating series of experiments by Babbage and Herschel with Arago's disc came tantalizingly near the answer.

'... we mounted a powerful compound horse-shoe magnet, capable of lifting 20 pounds [$\approx$ 9 kg], in such a manner as to receive rapid rotation about its axis of symmetry placed vertically, the line joining the poles being horizontal and the poles upwards. A circular disc of copper, 6 inches [≈ 15 cm] in diameter and 0.05 inch [≈ 1.3 mm] thick, was suspended centrally over it by a silk thread without torsion, just capable of supporting it. A sheet of paper properly stretched was interposed, and no sooner was the magnet set in rotation than the copper commenced revolving in the same direction at first slowly, but with a velocity gradually and steadily accelerating. The motion of the magnet being reversed, the velocity of the copper was gradually destroyed; it rested for an instant, and then immediately commenced revolving in the opposite direction...

'The rotation of the copper being performed with great regularity, it was evident that by noting the times of successive revolutions, we would acquire a precise and delicate measure of the intensity of the force urging it, provided we took care to neutralise the torsion of the suspending thread'.

So, Babbage and Herschel avoided the difficulty of Barlow's experiment by using Arago's method of generating 'rotational magnetism'. *Why was the 'sheet of paper properly stretched' needed?* They went on to test the effect of interposing different metal discs but this was found to be negligible except in the case of tinned iron. *This has, of course, a very good reason.*

The next step in the experiment gave a clue to the cause of the motion of Arago's disc and hence a chance to discover a new phenomenon. Discs of different materials were tested and compared with the copper one:

'[The method] consisted in securing each of the 10 inch [≈ 25 cm] discs... successively on the vertical axis of our machine... Giving them thus a rotation in their own planes, the azimuth compass... was placed on a convenient stand centrally over each at the same distance. The deviations observed, and the ratio of their sines to that of the deviations produced by... copper, were as follows:'

Material of revolving body	*Ratio of force to that of copper*
Copper	1.00
Zinc	0.90
Tin	0.47
Lead	0.25
Antimony	0.11
Bismuth	0.01
Wood	0.00

But this list is not an order of relative ease with which materials can be magnetized. *Inspect the list closely and suggest another property that it might represent. This is a critical clue!*

On occasions, a biographical approach and a case study might amount to the same thing, especially when some product of science is largely associated with the work of one scientist. Generally, however, in one the focus tends to be on the person while in the other it is on the product. Both approaches may, as in the example above, draw on scientific papers. Although their use is described as a separate approach in the next section, in practice these three may merge in highly integrated material.

Scientific papers

The use of scientific papers in school science teaching is probably not extensive. Their language and use of abstract mathematics render them unintelligible to younger pupils. Short, edited extracts from them and possibly their illustrations can give an air of authenticity to biographies and case studies. For example, pictures of Dalton's chemical symbols, illustrations from the voyage of the *Beagle* and Copernicus's so-called heliocentric system can vitalize second-hand prose. Older students, however, can gain something from well-chosen research papers (aims 5.0). Careful selection and some editing is usually necessary but, for example, the lucid account of the measurement of the half-life of 'thorium emanation' by Rutherford and Soddy is well within the grasp of sixth formers (Wyatt, 1981). Another useful way of overcoming the gap between sixth formers and reported research is to prepare annotated versions of papers relating to their course work. Some students may well derive valuable perspectives from their own reading of publications of a more popular nature such as, Capra's *The Turning Point*, Bronowski's *Ascent of Man* and Burke's *Connections* (aims 1.0, 3.0 and 5.0).

Dissection

It is my thesis that science teachers must make provision for the wider aims of science education. However, there is at least one practice which may be inimical to the achievement of those aims. There was a time when biology was definitely not the study of life. Only dead, dissected objects seemed worthy of study. Those that were alive were perfunctorily killed, pinned down, sliced up, then thrown away. Some teachers are now of the opinion that dissection has the potential to brutalize. If this is the case, then its role in school science must be seriously questioned, certainly in the formative years, since it would conflict with the fundamental aim of education, namely, to humanize. To add to this, there are also those who question the right of one species to use another in this way. Teachers need to decide whether they think dissection is an acceptable approach to teaching aspects of zoology (ASE, 1985; RSPCA, 1986; Vines, 1986).

But this is not to say that the subject of dissection should be avoided from the polemical point of view. Pupils can be given the opportunity to receive and air opinions on the benefits of dissection to mankind. This will usually lead to discussion on vivisection where Frederick Banting's work on diabetes using the rhesus monkey is a fairly clear instance of a benefit. Less clear is the use of animals to test cosmetics (aims 2.4 and 4.0; Lightner, 1974). In any case, the view of the issue is related to the details of the current world view (aims 3.0). J J Audubon killed the birds he painted in large numbers. Would a modern artist feel justified in doing the same?

Planning to achieve wider aims

Much of what has been said under this title in the previous chapter is also applicable here. Not all aims can or should be attempted on any one occasion. A sensible and realistic priority must be afforded to the teaching of science while balancing it with an honest attempt to achieve the wider aims. The opportunities which arise for this depend on the topic, the stage of development and the level of experience of the learner. Often, humanized and humanizing material can be used at several levels provided due allowance is made for age and experience. Younger children may benefit more from an approach which knits the material together and stitches it to what they already know. Older students may be able to make some of these links themselves and so a looser approach might be tried.

Generally, GCSE syllabuses do not offer guidance on teaching approaches as far as wider aims and relevant science teaching is concerned. Presumably, this area is seen as the prerogative and responsibility of the teacher who, knowing the students and the local situation, will adopt whatever methods are seen as being appropriate and likely to be effective. A notable exception is that of the Midland Examining Group in their electronics syllabus (1988) which discusses how social, economic and environmental implications might be approached. It suggests that they might either be taught 'formally' near the end of the course or taught 'informally' in parallel with the rest of the course. The latter approach is favoured on the grounds that it is more likely to tie the content closely to relevant contexts. The syllabus also lists six reasons for the impact of electronics on society: low cost, reliability, economy with rare resources, compactness, speed of operation and flexibility. Some nine consequences are added, for example: the replacement of effective mechanical devices with cheaper electronic ones, the creation of a throwaway technology, the automatic control of complex situations, and the possibilities for improving the quality of life for the disabled. Other syllabuses are not so helpful.

Finally, it should be added that, whatever the course and whatever the age, students are likely to benefit from variety in approach. Most of the material which has been referred to is readily available. Full details are given below. The reader's attention should also be drawn to the SATIS material of the ASE (1986b) and to the work of writers like Solomon (1980).

References

ASE (1985) *The Place of Animals in Education* ASE/Institute of Biology/ Universities Federation for Animal Welfare, Hatfield.

ASE (1986) *Science and Technology in Society* Association for Science Education, Hatfield.

Bates, E B (1980) The industry/education unit. *Education in Chemistry* **17**, 81–83.

Barratt, M S and Stanyard, T N (1979) The Davy bicentenary event. *School Science Review* **61**, 203–213.

Bartle, K D (1976) Polychlorobiphenyls – a case study in chemical pollution. *School Science Review* **57**, 265–275.

Borrows, P (1984) The Pimlico chemistry trail. *School Science Review* **66**, 221–233

Burton, W G (1979) Acceptability equations and case studies of three major disasters involving industrial chemicals. *School Science Review* **60**, 624–634.

Campbell, J M; Drummond, A (1977) The energy game. *School Science Review* **59**, 13–23.

Cherfas, J; Connor, S (1983) *How restless DNA was tamed. New Scientist* 13 October 1983, 78–79.

Collinge, E R (1972) *River pollution. School Science Review* **54**, 276–80.

Collinge, E R (1974) *Air pollution. School Science Review* **56**, 5–13.

Crawley, G M (1975) *Energy* Collier Macmillan, London.

Dessel, N F, Nehrich, R B; Voran, G I (1973) *Science and Human Destiny* McGraw-Hill, New York.

Douglas, A M (1986) Tigers in Western Australia? *New Scientist* 24 April 1986, 44–47.
Dowdeswell, W H (1981) The protein project. *Case Studies in Technology* BP Educational Service, London.
Dowdeswell, W H; Wells, J (1977) The Ridpest File. *Journal of Biological Education* **11**, 53–58.
Dussart, G B J (1981) The teaching game for applied hydrobiology. *Journal of Biological Education* **15**, 123–136
Eakin, R M (1975) *Great Scientists Speak Again* University of California Press, Berkeley.
Ehrenfeld, D (1986) Thirty million cheers for diversity. *New Scientist* 12 June 1986, 38–43.
Elliott, C (1978) Science. In Carson, S. McB. (ed.), *Environmental Education* Edward Arnold, London.
Fricker, J (1986) The waste tip that blew up a bungalow. *New Scientist* 10 April 1986, 24.
Gadd, P; Tyrell Smith, T S (1977) Project work in social biology at GCE Advanced Level. *Journal of Biological Education* **11**, 59–63.
Gee, B (1970) André Marie Ampère. *Physics Education* **5**, 359–369.
Greatorex, D; Lister, J M (1980) The organisation of project work in a mode 3 O-level chemistry course. *School Science Review* **61**, 427–440.
Harris, N D C (1981) The Rockwool Project. *Case Studies in Technology* BP Educational Service, London.
Harris, J; Osborne, J (1978) World energy supplies: the present use and future prospects. *School Science Review* **60**, 257–275.
Hecht, E (1980) *Physics in Perspective* Addison-Wesley, Reading, MA.
Heydenreich, L H *et al.* (1981) *Leonardo the Inventor* Hutchinson, London.
Jones, C E (1980) The chemistry of a Concorde. *School Science Review* **61**, 675–681.
Keller, E F (1986) One woman and her theory. *New Scientist*, 3 July 1986, 46–50.
Kempton, T; Allsop, T (1985) Science in society – a local development. *School Science Review* **67**, 223–230.
Lenihan, J (1979) *Science in Action* Institute of Physics, Bristol and London.
Lightner, J P (1974) Comment. *Journal of Biological Education* **8**, 247–248.
Lewis, J (1980) Industry's social role. *Education in Chemistry* **17**, 75
Lovelock, J (1986) Gaia: the world as living organism. *New Scientist* 18 December 1986, 25–28.
May, P F *et al.* (1980) Gypsum: a school industry science project. *School Science Review* **61**, 405–417.
Nellist, J (1980) Science lessons from industrial processes – Sunderland. *Education in Chemistry* **17**, 69–71.
Newton, D P (1980) Undercurrents, *Physics Education* **15**, 112–116.
Newton, D P (1986) The world view of Linnæus. *Journal of Biological Education* **20**, 175–178.
Pearce, F (1986) How to stop the greenhouse effect. *New Scientist* 18 September 1986, 29–30.

Reid, N (1980) Understanding chemical industry – teaching materials. *Education in Chemistry* **17**, 78–80.
RSPCA (1986) *Dissection* RSPCA Education Department, Horsham.
Searle, C E (1980) Smoking and disease – where should prevention start? *School Science Review* **62**, 19–36.
Scott, W A H (1981) *The Grangemouth Project* BP Educational, London.
Smith, J V (1986) The defence of the Earth. *New Scientist* 17 April 1986, 40–44.
Solomon, J (1980) The Siscon-in-Schools project. *Physics Education* **5**, 318–324.
Spencer, J (1977) Games and simulations for science teaching. *School Science Review* **58**, 397–413.
Strandh, S (1982) *Machines* Nordbok, Gothenberg.
Vines, G (1986) Experiments on animals: a balance of interests. *New Scientist* 24 April 1986, 26–27.
Wilkins, M (1986) Science, peace and life. *New Scientist* 13 March 1986, 48–49.
Wray, E V (1968) Nature trails as a teaching aid. *Journal of Biological Education* **2**, 21–38.
Wray, J D (1976) What does conservation mean? *Journal of Biological Education* **10**, 115.
Wyatt, H V (1981) Using the classics. *Journal of Biological Education* **15**, 79–80
Wynne, B (1979) C G Barkla and the J phenomenon. *Physics Education* **14**, 52–55.

CHAPTER 6

SOME CAUTIONARY NOTES

Over-gleaning

Having painstakingly searched for, selected and edited material, there is an understandable tendency to want to wring every last drop from it. A lesson may well be replete with opportunity for teaching about and through science, but not all of it will be appropriate. For young children, it might be the material relating to themselves, their family, their friends and the local community which is appropriate. For older students, it might be that which relates to the nature of science and its relationships with society. What is important is what the learner can relate to, finds interesting and can integrate with his or her experience. At times, an opportunity might pass without a flicker of interest on the part of the pupil. At other times, the teacher might seize on a ready interest and turn it towards the wider aims. To begin with, the teacher will need to explore approaches, find out what can be achieved and acquire some new skills. At this stage, a surfeit of opportunity is a positive advantage.

It is doubtful if a given aim could be achieved fully by one brief contact with humanized material. Children will need to see the view from many sides and in a number of guises before their horizons are genuinely broadened. Even on a good day, when you have a classful of budding philosophers, it is probably better to channel that mental energy towards one well-defined goal rather than dissipate it across the width of the framework and risk mental indigestion. A grand scheme of science and its ramifications could not be inculcated in a single science lesson; it must be carefully constructed over a period of time. Jennings (1980) warns that, 'It is important for teachers to set limited, obtainable goals rather than to subscribe to aims which are so ambitious that a substantial proportion of pupils appear to fail'.

Questions of balance

For those who take readily to providing for the wider aims of science teaching, it is easy to let their enthusiasm get out of hand. The science in the lesson is lost in history, philosophy, current affairs, sociology and psychology. Their students may know all about the relevance and ramifications of science at the expense of the science itself. It is not necessary for all science teaching to support general education: knowledge of science is useful and can have a vocational aspect (Crary, 1969).

It may be quite the reverse for those who have no taste for wider aims. Merely to dabble shallowly in humanized science teaching may do little to combat over-optimistic scientism, a belief that science provides the only valid model for living and for solving human problems. It similarly leaves untouched the anti-science attitudes at the other extreme (Holton, 1976; Solomon, 1980). The capacities to be developed in the pupils need preparation, thought and planning and are unlikely to arise from off-the-cuff remarks about today's headlines and apocryphal tales about Newton, Dalton or Darwin. The Secondary Examinations Council recommends that about ten per cent of the science teaching time be given to the social, environmental and economic aspects of science. In a class's weekly allocation of two hours of science this amounts to a twelve minute slot. If such aims are to be achieved in a GCSE course, and I believe a serious attempt must be made to do so, then there is no alternative but to use an economical, highly integrated approach in which the teaching of, about and through science are indissolubly fused into one. This needs planning. Should the same percentage be recommended for sixth formers, it would amount to about a half-an-hour a week. Since it would be reasonable to expect some of their non-class time to supplement it, this is potentially useful.

As well as maintaining an appropriate weighting of products, processes and people, care is needed to ensure a reasonable balance between the wider aims. Taking into account the stage of development and ability of the children, those aims considered appropriate should be suitably represented in the planning over a period of time. At an even finer level of preparation, care needs to be taken to ensure an appropriate balance within material designed to achieve a given objective. The detrimental effects of technology, for example, are more newsworthy than its beneficial effects. Big issues largely tend to be centred on the former. Is that because they are more common or is it a reflection of selection by media? If it is the latter, then there should be balancing material relating to the beneficial effects provided by the teacher. Those effects which fall into neither group and are of marginal value probably receive least publicity of all.

When encouraging a sense of the tentativeness of theories we, in the wealth of our knowledge and experience, know that there are many kinds of theories and many levels of certainty. We also know of the disagreements among philosophers regarding the nature of science. Children know none of this. We must be careful to keep it in proportion or they may leave thinking that tomorrow there will be no molecules in chemistry, no electrons in physics and no genes in biology. Science will become guesswork – trivial and irrelevant. While the long-term impermanence of a paradigm needs to be appreciated, it is the responsibility of the teacher to pass on the substance and authority of the prevailing paradigm (Biggins and Henderson, 1978).

More difficult to detect in our material is an unconscious bias in selection and presentation. I assume here that no professional teacher would knowingly use humanized science teaching to propagate a particular social theory or belief and would be concerned to minimize the political indoctrination of children. Rather, it is the exemplars, issues and topics we choose or prefer to use in our work that, taken together, reflect our own value judgements and so prejudice the outcome of the lesson. Bias of a political nature is not the only concern. Attitude studies have shown that girls are concerned about the social implications of science (Johnson and Murphy, 1986). They are attracted to subjects in which such concerns are discussed. Our zeal to recruit asexually and widely must not overcome a scrupulous attempt at honesty in the picture of science we present. Sharing views and materials with a sympathetic colleague of a different persuasion can be a valuable guard against bias.

Limitations

Some approaches have a great potential for humanizing science teaching and showing the relevance of science. Historical material, for example, is so rich in opportunity that it seems it could be used to achieve any of the wider aims (Fig. 6.1). Readily available and lending itself well to the exposition of the science itself, it can offer both an economical and effective way of embedding science in coherent, human contexts. But no one approach is a panacea for the present shortcomings of science teaching. The

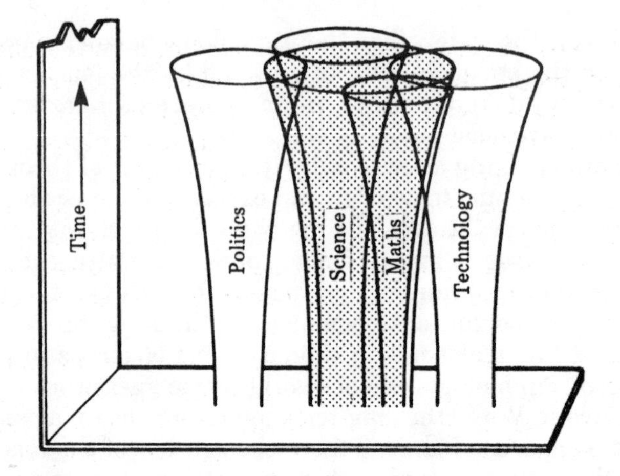

Figure 6.1 *Science has grown, developed and touched on an increasing number of other human activities. The areas of interaction change with time. Reid's interface diagram (Fig. 5.4) is a horizontal slice through the figure, representing today's areas of interaction.*

twentieth century is unique. What is happening now in science and society may have no true parallel in what has happened before. Consequently, the historical material we choose may give an inappropriate image of the progress of science (Whitaker, 1979). The main aim of such an approach is not usually to teach the history of science. For example, topics are chosen because they illuminate aspects of human nature and the impact of science and technology on society. As a consequence, we must take care not to be cavalier with history. In our desire to present an orderly, trim picture to suit our ends, we must avoid rewriting history, restructuring and grossly oversimplifying the complex course of events, and judging the past with the attitudes and values of the present. There is also a danger that we might import the attitudes and values of the past into the present. Men monopolized science in the past and still do, especially in the physical sciences (Kahle, 1985). Historical exemplar material will probably reflect this and could reinforce unwanted role preferences in the young unless remedial action is taken. Similarly, the institutionalization and team-based approach of modern science is not well-illustrated by the work of the solitary, gifted amateur of the early days and 'biographies' of research teams, in a form suitable for children, are not very common.

I doubt if any single approach is free from limitations of one kind or another. Topical issues, for example, relate to modern science; men and women, working in teams on problems which could radically affect the lives of the pupils. But what is newsworthy tends to follow popular emotion and sentimentality. It clusters around what makes 'good copy'. Topical issues seldom relate to many of the wider aims and the topic is out of the control of the teacher. Furthermore, background information is dispersed and less well digested than that which has been under the scrutiny of an historian.

The lesson is that the approaches are often powerful but their shortcomings must be recognized. Consequently, the armoury should contain a number of approaches which complement one another and offer support in the areas of weakness which any one possesses.

Circumspection

Experience tells most science teachers that some topics need to be approached with care. Accounts of incurable diseases, aspects of nuclear war, population control and vivisection, for example, may offend sensibilities, horrify, insult religious beliefs or release strong emotions because of some personal circumstance. An adolescent of fifteen or sixteen years of age, close to someone dying of cancer, for example, may be unable to confine his emotions in a lesson on the medical uses of gamma rays. Very young children, on the other hand, commonly do not attach great significance to death but we are rightly concerned not to cause them undue worry or mental disturbance. Death as a fact and concept takes time to come to terms with. Developing minds are vulnerable as half-digested concepts are distorted and loom large, even in late adolescence. The experienced teacher knows this and unconsciously tunes the approach to excite just the right measure of involvement, concern and expectation.

Assessment and evaluation

The pupils are actively engaged in their science. They show interest, are keen to try out ideas and seem able to relate what they do to themselves and to others. All seems to be going well. In such circumstances, it is easy to believe that the goals we set are being achieved. Perhaps they are – the conditions certainly seem right – but some assessment of progress would be helpful. The advantage of such an assessment is that:

- it helps to determine a pupil's progress towards a goal and facilitates decisions about provision for further progress;
- it helps to diagnose areas of weakness or difficulty for a pupil so that remedial action can be appropriate;
- it might be used as an indicator of a pupil's level of achievement, absolute or relative;
- it shows the pupil that the aims are valued by the teacher and, by implication, by society at large;
- it can indicate the value of an approach or topic for achieving particular objectives.

An understanding of what is meant by *progress* is a pre-requisite of such an assessment. It must be defined in such a way that it is meaningful, relevant and measurable. This means that the objectives of a lesson, topic, block of work or course must be clearly stated. It helps to ask, 'At the end of this lesson/topic/... I want the pupil to...' The answer should be something which can be observed, be it knowledge, skills and processes or attitudes. For example:

A '...I want the pupil to be able to state two effects of pollution observed in stream X';
B '...I want the pupil to be able to make a value judgement of the utility of Y and support that judgement with at least one rational reason';
C '...I want the pupil to be able to list some alternative solutions to problem Z and to indicate those which involve questions of morality'.

Progress towards the objectives is then defined in terms of such statements. For **A** it might be:

1 unable to state any effects of pollution;
2 able to state an effect with prompting and assistance;
3 able to state an effect without prompting... etc..

For **B**:

1 unable to make any value judgement;
2 can make a simple value judgement (useful/useless) but unable to express a rational justification of it;
3 can make a simple value judgement and can express a rational justification of it;
4 can make a graded value judgement... etc.

For **C**:

1 unable to list any solutions;
2 able to give one solution;
3 able to give alternative solutions;

and

1 unable to identify the solutions involving questions of morality;
2 able to identify some of the solutions involving questions of morality... etc.

Often, information on progress is collected informally, as a teacher observes, listens to and discusses work with groups and individuals. Provided that each pupil is given adequate opportunities to show his progress, this form of monitoring is useful. Other sources of information are the devices and models constructed by the pupils, their drawings, conclusions, verbal and written accounts, special assessment tasks and answers to written questions. Each has advantages and disadvantages at different stages of education and with different levels of ability. When a pupil profile of the kind; *usually can, sometimes can, never can* is required, then assessment must be fairly regular if this is to be meaningful. Using all the sources of information available helps to alleviate the work and prevent assessment from dominating a course.

The subject of assessment and evaluation is a large and often difficult one and is the focus of much research. For the primary school level, teachers might find the work of Harlen (1985) useful, while at the secondary level that of Ward (1980) and Sutton *et al.* (1986) might be of interest.

Changing emphases

As time passes, the role of science in our lives may change. Science education must reflect this change in the weightings it gives to the various aspects of the framework described in Chapter 3. At the same time, teachers must continue to develop an awareness of the breadth of that role, whatever the emphasis is to be. The relevance of science is not confined to our personal interests and to transient issues. Science education must help to prepare our future citizens for their future.

References

Biggins, D R; Henderson, I (1978) What is science teaching for? *Physics Education* **13**, 438–441.

Crary, R W (1969) *Humanizing the School: Curriculum Development and Theory* A A Knopf, New York.

Harlen, W (1985) Assessment and record keeping as part of teaching primary school science. In *New trends in primary school science education*, Vol. 1, UNESCO.

Holton, G (1976) The Project Physics course. *Physics Education* **11**, 330–335.

Jennings, A (1980) Comment: alternatives for school science. *Journal of Biological Education* **14**, 193.

Johnson, S; Murphy, P (1986) *Girls and Physics* APU/Department of Education and Science, London.

Kahle, J B (1985) *Women in Science: A Report from the Field* Falmer Press, London.

Solomon, J (1980) The Siscon-in-Schools project. *Physics Education* **5**, 318–324.

Sutton, R *et al.* (1986) *Assessment in Secondary Schools: The Manchester Experience* Longman, York.

Ward, C (1980) *Designing a Scheme of Assessment* Stanley Thornes, Cheltenham.

Whitaker, M A B (1979) History and quasi-history in physics education, *Physics Education* **14**, 108–112 and 239–242.

CHAPTER 7

CHANGING THE WAY SCHOOL SCIENCE IS TAUGHT

Shrinking from wider aims

Science teachers have tended to shrink from the wider aims of science education. A survey of physics teachers in 1974 showed little enthusiasm for them. The majority thought that the place of the history of science and its social effects was in the sixth form General Studies course and not in the physics lesson. It was accepted that engineering applications of physics should be covered in a course on physics but the majority thought they should not appear on examination papers (Holley, 1974). Jennings (1983), referring to biology teaching, says that courses fail to set man in context in the living world and that course diversity is more apparent than real. In other words, as Jungwirth (1980) has expressed it, the biology curriculum has tended to treat *Homo sapiens* to the total exclusion of *Homo anthropos* – man in society.

Yet the wider aims described here are not new. Since the early days of science education, teachers have been urged to widen their teaching for one reason or another. Obviously, these exhortations have been largely ignored or overwhelmed by other priorities. However, since the Thomson Report of 1918, enormous social, environmental and economic changes seem to have made the necessity for wider aims more pressing. So why do science teachers behave in this way? No doubt, conservatism plays a part. By and large, people like to keep to secure, familiar patterns of behaviour, and science teachers are no exception. Inertia in teaching is far greater than is often realized. As Waring (1983) points out, 'The history of science education is drawn mainly from the bulkier data of interest group rhetoric and policy-making. It does not reflect the impact of ideas in the classroom'. Change in educational practice is also often more apparent than real.

Teacher specialism is another reason for resistance to change, at least in the secondary school. Science teachers are not usually 'generalists'. They are experts in a body of knowledge in great demand and form an élite in the teaching force. Anything which seems to dilute this specialism is seen as a threat to be described as 'soft' science.

Another contributory factor is that science has tended to be taught in the way it was thought to be practised, that is, dehumanized. Didactic and fact-oriented teaching styles have developed which achieve this efficiently. The achievement of wider aims would often need approaches and skills that tend to be unfamiliar to the science teacher. This may be compounded by decades of self-selection in which those who become science teachers are happier with dehumanized approaches.

In such a climate, there may also tend to be a shortage of suitable teaching material or little diversity in what is available. Much textual material concentrates on being a surrogate teacher of the products of science. They duplicate the teacher's efforts. There is a real role for textbook writers who will prepare material with which to supplement the conventional lesson and extend it into the areas described here. The book and teacher would become mutually supporting partners who, together, would provide for all the aims of science education (Newton, 1984). Writers might also explore ways of presenting integrated approaches so that these aims are achieved economically as far as time is concerned. Some material is becoming available but it needs to be assessed against the background of the framework described in Chapter 3. Teachers will have to check that its aims are appropriate and be aware of neglected yet suitable areas which they will still need to develop.

Examination pressures are another factor in causing the neglect of wider aims. Since the achievement of wider aims is seldom, if ever, assessed, they receive little more than lip-service. After all, to do otherwise would be to waste valuable time which could be used to ensure examination success. It is examination success that parents and headteachers want. If the role of formal examinations was to become less important it would reduce this kind of pressure on the teacher. Students need not then be penalized for learning about and through science. Of course, a reduction in pressure does not, in itself, imply that humanized science teaching will flourish in our secondary schools. Too many other forces are at work to be sure of that. Science teachers must be inclined to take the opportunity offered to them. On the other hand, if external examinations were increased in frequency and taken by young children, as has been suggested, then the temptation for the teacher to treat science narrowly could increase. If necessary, ways should be found to assess the achievement of the wider aims of science teaching within examinations.

The role of higher education, teacher training and in-service courses

It would undoubtedly help if many university science courses did not allow undergraduates to become divorced from study about science. The history and philosophy of science, for example, would be very useful for the would-be teacher. Myths about scientists and the practice of science might then be dispelled and those who turn to teaching science would be better placed to judge what is required. But university science departments would probably argue that this is not their concern.

In the absence of an adequate background for achieving wider aims, it falls to the lot of teacher trainers to make up that deficiency. It may be necessary to prescribe and support reading programmes for student teachers to fill in the gaps in their science education. It is equally important that novice science teachers be introduced to the methods of teaching more commonly found in the humanities. Their school visits should not be confined to science departments. Observation of teaching skills and methods in other disciplines is needed. Student teachers need to develop the capacity to encourage and lead discussion into those areas where answers are not simply right or wrong.

In-service courses to achieve the same end would be a useful aid to change. At present, advisory teachers in many parts of the country are giving support and courses for primary teachers who have to teach science. This work is largely concerned with initiating and nurturing science teaching in the primary school. There is a danger that it will follow the pattern already found in the secondary school, namely, focusing exclusively on products and processes. This practice has been described here as deficient. While these should be afforded priority, the attention of teachers should be drawn to that essential third component; people.

GCSE requirements include the achievement of wider aims. It might be useful for secondary science teachers to receive some guidance in what this entails, how to prepare and assess material and how to humanize their science teaching. At both the primary and secondary level, school-based in-service work is now developing. This is another opportunity for improving the quality of science teaching in both phases of education. Primary and secondary school links are also being established in many areas and science teachers in the two phases may well find it useful to compare ideas and practices at these meetings.

In conclusion

Science will occupy a significant part of every schoolchild's education. Only a tiny proportion will become scientists, so the main thrust of science education will be towards increasing scientific literacy. According to O'Hearn (1976), scientific literacy has four components: scientific knowledge, skills and processes associated with science, the nature of science and the social and cultural implications of science. Teaching about products and processes may achieve literacy in the first two and, indirectly, some implicit understanding of the third. The other ingredient, people, is needed to complete the task. Jennings (1980) reminds us that, 'For too long, school science has tended to neglect the imaginative, creative dimension of science and its contribution to our culture'.

Society in the year 2000 will need to understand discussions involving science and technology as they are communicated by the media (Jacobson and Bergman, 1980) and to participate in those discussions. It will need to make value judgements and take decisions affecting the quality of life. People will, as always, have aspirations but they will need to act responsibly and be able to relate to and respect others (Palm, 1974). More and more these needs, concerns, decisions and actions will involve an aspect of science and technology. H G Wells has expressed this trend well. 'Human history becomes more and more a race between education and catastrophe'.

The people of the year 2000 are in our schools now.

References

Holley, B J (1974) *A-level Syllabus Studies: History and Physics* Schools Council/Macmillan, London.

Jacobson, W J; Bergman, A B (1980) *Science for Children* Prentice-Hall, New Jersey.

Jennings, A (1980) Comment: alternatives for school science. *Journal of Biological Education* **14**, 193.

Jennings, A (1983) Biological education – the end of the dinosaur era? *Journal of Biological Education* **17**, 298–302.

Jungwirth, E (1980) Some biology/social science interfaces and the teaching of biology. *Journal of Biological Education* **14**, 339–344.

Newton, D P (1984) Textbooks in science teaching. *School Science Review* **66**, 388–391.

O'Hearn, G T (1976) Science literacy and alternative futures *Science Education* **60**, 103–114.
Palm, A (1974) Human values in science. *BioScience* **24**, 657–659.
Waring, M (1983) The roots of curriculum inertia. *Journal of Biological Education* **17**, 273–274.

CHAPTER 8

IDEAS AND INFORMATION

Teaching materials

PRIMARY/MIDDLE SCHOOL

Few commercial schemes seem to relate science explicitly, and to any great degree, to the needs of people. *Scienceworld* (Longman, 1986) does relate science to the child, and to friends and family to some extent. *Young Scientist Investigates* (Oxford University Press, 1986) and *Reading about Science* (Heinemann Educational, 1982) also point out the relevance of science but with more emphasis on people in general rather than on those closest to the child.

The teacher would need to supplement most primary school science material if the relevance of the content is to be made explicit in the way described here. *Footsteps into Science* (Stanley Thornes, 1987) is a set of workcards written with the intention of producing a humanized primary/middle school science teaching resource. It could be used alone or could supplement existing courses without great expense.

LOWER SECONDARY SCHOOL

The situation is similar at this level, much of the commercially produced material being dehumanized. A notable exception is *Nuffield Science 11 to 13* (Longman, 1987). To a lesser extent are *Warwick Process Science* (Ashford Press, 1987) and *Science in View* (Oxford University Press, 1986). However, as packages expected to achieve more of the aims of science teaching than described here, they are very different. The first two provide details of practical activities, the last is intended to supplement or to stimulate a practical course. Each has strengths and weaknesses, and has been written from a particular view of science and, therefore, of science education.

UPPER SECONDARY SCHOOL

Generally, GCSE texts are not particularly humanized in the way described here but some useful supplementary material is available. For example, there are the *Science and Technology in Society (SATIS)* units (Association for Science Education, 1986 *et seq.*), the *Experimenting with Industry* series (Standing Conference on Schools' Science and Technology/ASE, 1986 *et seq.*), and the *Finding out . . .* and the *Physics Plus* series (Hobson).

In addition to these booklets, leaflets and charts, there are some class textbooks written as course supplements, for example, the *Reading about Physics, . . .Chemistry, . . Biology* series (Heinemann, 1986). Some course books claim to make their exposition relevant but they are often disappointing. *Chemistry in Use* by R Jackson (Pitman, 1984) is an exception.

There is a need to be aware of the deficiencies of these materials. Some point out only the utility of science through technological applications. Such material may still need to be supplemented.

For the teacher, there is *Better Science: A Directory of Resources* (Heinemann Educational/ASE, 1987). Of course, there are books which will provide useful data and information which are not, in themselves, suitable for class use, for example: *Energy* by G M Crawley (Macmillan, New York, 1975) which describes the science of energy resources.

SIXTH FORM

Textbooks for the Sixth Form are probably no better or worse than others as far as explicit relevance is concerned but the greater knowledge, experience and relative maturity of the Sixth Former allows a greater range of supplementary material to be used.

As at the lower level, there are commercially produced booklets, like the *Science Support Series* (Hobsons) which demonstrate 'the applications of pure science in the working world'. There are also a few texts which humanize science, like *Project Physics* (Holt, Rinehart and Winston, New York, 1975) and, to a lesser extent, *Higher Physics* by J Jardine (Heinemann, 1983). However, also available are edited and annotated collections of scientific papers such as, *Classical Scientific Papers – Physics* by S Wright (Mills and Boon, 1964), *The Concept of the Atom* by C Butler *et al.* (Heinemann, Australia, 1970) and *Great Scientific Experiments* by R Harré (Oxford University Press, 1981).

Not to be overlooked as sources of information and ideas, are the General Studies materials like *Science in Society* (Heinemann/ASE) and *Science in a Social Context* (ASE). The Cambridge, Oxford and Southern School Examination Council, through their examination boards produce a Science in Society syllabus for the AS level of the GCE which includes useful specimen material and a list of resources.

While these materials indicate relevance mainly through applications of science, the history of science or by reference to its impact on society, John Lenihan's *Science in Action* (Institute of Physics, 1979) offers amusing anecdotes about the people who do the science and dissolves the artificial boundary between the scientist and the rest of society.

Journals like the *School Science Review* and *New Scientist* will also be found useful as sources of ideas, topics and information.

Journals and associations

The journals listed here should be available in education libraries for consultation. Journal titles are shown in italics.

Association for Science Education, College Lane, Hatfield, Hertfordshire AL10 9AA
Holds conferences, meetings and workshops for the advancement and development of science education in both the primary and secondary school. Publishes a number of journals:
Education in Science
The official means of communication of the ASE;
Primary Science Review
Primary education;
School Science Review
Mainly secondary education but with general articles which span the primary and secondary phases.

British Association for the Advancement of Science, Fortress House, 23 Savile Row, London W1X 1AB
Offers an Awards for Young Investigators scheme and has a section for Young Scientists (BAYS).

Child Education, Scholastic Publications Ltd, Marlborough House, Holly Walk, Leamington Spa, Warwickshire CV32 4LS
Mainly for infant education.

Council for Environmental Education, School of Education, University of Reading, London Road, Reading RG1 5AQ
Publishes leaflets on aspects of environmental education and a regular bulletin sheet.

Education in Chemistry, Royal Society of Chemistry, Burlington House, London W1V 0BN
Secondary education and higher.

Journal of Biological Education, Institute of Biology, 20 Queensberry Place, London SW7 2DZ
Secondary education and higher.

Junior Education, Scholastic Publications Ltd, Marlborough House, Holly Walk, Leamington Spa, Warwickshire CV32 4LS
Mainly for junior education.

New Scientist, New Science Publication, Holborn Publishing Group, Commonwealth House, 1–19 New Oxford Street, London WC1A 1NG
Articles on science for the general reader with a science background.

National Association for Environmental Education, 27 Queens Road, Hertford SG13 8AZ
Publishes:
Environmental Education
Primary and secondary environmental education.

Physics Education, Institute of Physics Publishing Ltd, Techno House, Redcliffe Way, Bristol BS1 6NX
Secondary education and higher.

STEAM, ICI Science Teachers' Magazine, ICI Educational Publications, PO Box 50, Wetherby, West Yorkshire LS23 7EZ
Articles on science and engineering for the secondary school.

Teaching Science, School Natural Science Society, 9 Killington Drive, Kendal, Cumbria LA9 7NY
Primary and lower secondary education.

The Green Teacher, Open Road Co-op, 3a Barker Lane, Micklegate, York YO1 1JR
Articles on environmental issues for the primary and secondary school.

The Science Teacher, 1742 Connecticut Avenue NW, Washington, D.C. 20009
Articles on science education in the American junior and senior high school, but often of interest to the British science teacher.

Technology in Education, B&S Publications, Draggetto Court, Chapel Hay Lane, Churchdown, Gloucester GL3 2ET
Articles on craft and technology, including electronics, in the secondary school.

Useful addresses

Association for Science Education
College Lane
Hatfield
Hertfordshire AL10 9AA

Association for the Conservation of Energy (ACE)
9 Sherlock Mews
London W1M 3RH

BP Educational Service
PO Box 5
Wetherby
West Yorkshire LS23 7EH

British Gas Education Service
Room 707A
British Gas
326 High Holborn
London WC1V 7PT

British Museum (Natural History)
Publication Sales
Cromwell Road
London SW7 5BD

British Rail Education Service
PO Box 10
Wetherby LS23 7EL

British Telecom Education Service
81 Newgate Street
London EC1A 7AJ

British Trust for Conservation Volunteers
36 St Mary's Street
Wallingford
Oxfordshire OX10 0UE

Conservation Trust
c/o George Palmer School
Northumberland Avenue
Reading
Berkshire RG2 7PW

Council for Environmental Conservation
Zoological Gardens
Regents Park
London NW1 4RY

Department of Education and Science (DES)
Elizabeth House
York Road
London SE1 7PH

Department of the Environment
Romney House
43 Marsham Street
London SW1P 3PY

Education Service of the Plastics and Rubber Institute
Department of Creative Design
Loughborough University
Loughborough
Leicestershire LE11 3TU

Electricity Council
(Understanding Electricity)
30 Millbank
London SW1P 4RD

Energy Efficiency Office
Department of Energy
Information Division
Thames House South
Millbank
London SW1P 4QJ

Flora and Fauna Preservation Society
Stirchley Grange Environmental Interpretation Centre
Stirchley
Telford
Shropshire TF3 1DY

Friends of the Earth
377 City Road
London W1M 8DQ

Health Education Council
78 New Oxford Street
London WC1A 1AH

International Union for Conservation of Nature and Natural Resources
Avenue de Mont-Blanc
CH–1196 Gland
Switzerland

Institute for Earth Education
c/o Stewart Anthony
6 Wicklands Road
Hunsdon
Ware
Hertfordshire

Jersey Wildlife Preservation Trust
Les Augres Manor
Trinity
Jersey
Channel Islands

Marine Conservation Society
4 Gloucester Road
Ross-on-Wye
Herefordshire HR9 5BU

National Society for Clean Air
136 North Street
Brighton BN1 1RG

Nature Conservancy Council
19–20 Belgrave Square
London SW1X 5PY

Oxford and Cambridge Schools Examination Board
Elsfield Way
Oxford OX2 8EP
or
10 Trumpington Street
Cambridge CB2 1QB

Royal Society for Nature Conservation
The Green
Nettleham
Lincoln LN2 2NR

Royal Society for the Protection of Birds
Head of Education
The Lodge
Sandy
Bedfordshire SG19 2DL

RSPCA
Manor House
The Causeway
Horsham
Sussex RH12 1HG

Science Museum
Exhibition Road
South Kensington
London SW7 2DD

Shell Education Service
Shell UK Limited
Shell-Mex House
Strand
London WC2 0DZ

Southern Universities' Joint
Board for School Examinations
Cotham Road
Bristol BS6 6DD

Standing Conference on Schools'
Science and Technology (SCSST)
1 Birdcage Walk
London SW1H 9JJ

Town and Country Planning
Association Education Unit
17 Carlton House Terrace
London SW1X 8QN

United Kingdom Atomic Energy
Authority
(Information Services Branch)
11 Charles II Street
London SW1Y 4QP

United Nations Educational,
Scientific and Cultural
Organisation (UNESCO)
UNESCO House
7 Place de Fontenoy
75700 Paris
France

University of Cambridge Local
Examinations Syndicate
Syndicate Buildings
1 Hills Road
Cambridge CB1 2EU

Woodland Trust
Autumn Park
Dysart Road
Grantham
Lincolnshire NG31 6LL

Wildfowl Trust
Gatehouse
Slimbridge
Gloucestershire GL2 7PT

World Health Organisation
1211 Geneva 27
Switzerland

World Information Service on
Energy
52 Acre Lane
London SW2 5SP

World Society for the Protection of
Animals
106 Jermyn Street
London SW1Y 6EE

Worldwide Wildlife Fund
Panda House
11–13 Ockford Road
Godalming
Surrey GU7 1QU

AUTHOR AND REPORT INDEX

SUBJECT INDEX